OUT OF BOUNDS

Innovation and Change in Law Enforcement Intelligence Analysis

Deborah Osborne, Research Fellow

JMIC Press

Joint Military Intelligence College

Washington, DC

March 2006

Table of Contents

Acknowledgments

This book owes its existence to the many analysts and experts who were willing to share their stories with me. I recognize and appreciate that some participants were uncertain about the need to bridge the distance between analytical fields and missions in law enforcement, but nevertheless were willing to go along with my theory that this requirement is real by participating in the interview process. I thank them for their confidence in my work and their gracious gift of time and ideas.

The command staff of the City of Buffalo Police Department was also supportive of this work. I especially thank my immediate supervisor in the chain of command, Captain Mark Makowski, whose leadership has offered me the freedom to explore and create beyond the usual municipal civil service position. The training I have received as an analyst and the networking I have been encouraged to do in my position has made this book possible.

A special note of thanks to my father, Kenneth Patten, who encouraged me to look for the answers to my questions and always had faith in my abilities to do so.

Above and beyond them, this book exists because of the generous invitation and continual support of Dr. Russell Swenson, whose proposal to engage in this research project led to a most interesting year of inquiry and discovery.

Deborah Osborne
November 2005

Foreword

In the novel and on the movie screen, the suave detective and the hard-bitten-but-sensitive street cop get the glory. But behind the scenes in the real world, a crucial foundation of good police work is the collection, assimilation, analysis, and communication of information about events, places and people. Crime and intelligence analysis is the back-office process that frequently underlies the solved crime, the ameliorated problem, and the effective prevention strategy.

Deborah Osborne's *Out of Bounds: Innovation and Change in Law Enforcement Intelligence Analysis* addresses the changing nature and role of analysis in policing. Osborne's examination, though, focuses not only on the analytical process, but on the analysts — critical actors who function with relative anonymity.

Osborne employs a provocative method of study: appreciative inquiry. In essence, analysts tell their stories: what motivates them, what successes they have enjoyed, what processes have worked well for them, how they see the future. A picture emerges of women and men who have great passion for their work, and who make tremendous contributions to solving crimes, interrupting crime patterns, apprehending criminals, and even preventing crime. By studying what works, the appreciative inquiry process draws out the themes that characterize these successes: innovative thinking; creative problem solving; intra-agency teamwork; collaboration and information sharing among agencies.

One of the more significant traits uncovered among state and local agency crime and intelligence analysts is an overwhelming agility. These people are quick on their intellectual feet. They constantly adapt, try multiple approaches, quickly adopt technologies or methodologies they find helpful, cultivate allies and complementary partners across organizational boundaries, and find ways to overcome impediments. They can describe specific results achieved with great clarity, and can define their own contribution to the successes. This stands in bold contrast to the fuzzy goals and bureaucratic doublespeak that seem to characterize federal intelligence agencies. The locals appear not only to have their act together, but to be quite adept at leveraging information, technology, and people to achieve results. Of particular note, local analysts who function in a patchwork of jurisdictional overlaps and adjacencies, with divergent governing bodies and widely varying information systems, have found remarkably effective ways of bridging these potential divides and collaborating effectively. The Feds could learn a lot from the locals in this regard.

The description of the appreciative inquiry process itself is an intriguing aspect of the book. Although appreciative inquiry has rarely if ever been applied to the criminal justice process, Osborne carefully explains its purpose and demonstrates its value in this area. The strength of this approach becomes evident in the book as the themes unfold in the words of a diverse group of analysts working in vastly different circumstances. One can easily see appreciative inquiry's potential for other applications in law enforcement and beyond.

Tom Casady, Chief of Police, Lincoln, Nebraska

Commentaries

Deborah Osborne asserts the following truisms about the state of intelligence analysis in U.S. Law enforcement: that the information infrastructure and data quality found in most agencies does not support analysis; that a lack of understanding of what analysis can provide continues to hold us back; and that civilian analysts still have difficulty being accepted as professionals. This is all common knowledge among law enforcement analysts as well as police officers in a position to know, but still needs to be continuously hammered home. Although her informants offer useful insights, a broader array of participants would have provided even more depth in her analysis of the environment.

One of the most valuable conclusions reached by Ms. Osborne is that law enforcement analysts are motivated by the same intangibles regardless of what level of government they serve: they want to feel useful, challenged, appreciated for their skills and recognized as part of a team.

Our tendency in this country to continuously collect information for the sake of collecting it, and disregarding any criminal information which does not support arrest or criminal charges is a terrible waste of resources, and will continue to compromise our ability to keep this country and our communities safe. I am happy to see that the author highlights examples from outside the U.S., illustrating that there are more successful models of police intelligence (Royal Canadian Mounted Police, Northern Ireland's Analysis Centre) to which we can aspire.

Lisa Palmieri, President, International Association
of Law Enforcement Intelligence Analysts

Ms. Osborne's book offers a fresh assessment of what ails the Intelligence Community. While the powers that be — that is, federal intelligence agencies and the Department of Homeland Security — continue to struggle with intelligence sharing and an intelligence model that fails to recognize the value of local intelligence, this book offers a number of insightful observations that, if seriously considered, could make significant improvements in fighting terrorism. Her methodological approach; namely, Appreciative Inquiry, lends itself to creative ideas about improving the business of intelligence and crime analysis. This book focuses on a number of fresh ideas about "what works"— See Chapters 3-8. This book demonstrates the power of being open-minded and thinking critically about how we go about intelligence gathering, analyzing crime and criminals; and how we might collaborate better to counter terrorism and crime alike. This book brings together a plethora of ideas and viewpoints, and is a must read for analysts and analysis unit managers, as well as public safety CEOs.

Noah Fritz, President, International Association of Crime Analysts

Acronyms

AI	Appreciative Inquiry
BJA	Bureau of Justice Assistance
BJS	Bureau of Justice Statistics
CA	Crime Analysis
CALEA	Commission for the Accreditation of Law Enforcement Agencies
CDX	Counterdrug Intelligence Executive Secretariat
CICC	Criminal Intelligence Coordinating Council
CLEAR	Citizen Law Enforcement Analysis and Reporting (Chicago)
COP	Community-Oriented Policing
COPS	Community-Oriented Policing Services
COPPS	Community-Oriented Policing and Problem Solving
CPTED	Crime Prevention Through Environmental Design
DEA	U.S. Drug Enforcement Administration
DHS	U.S. Department of Homeland Security
DIA	Defense Intelligence Agency
DOJ	U.S. Department of Justice
FBI	Federal Bureau of Investigation
FDLE	Florida Department of Law Enforcement
GIS	Geographic Information Systems
GISWG	Global Infrastructure/Standards Working Group
GIWG	Global Intelligence Working Group
GSWG	Global Security Working Group
HIDTA	High Intensity Drug Trafficking Areas
IACA	International Association of Crime Analysts
IACP	International Association of Chiefs of Police
IALEIA	International Association of Law Enforcement Intelligence Analysts
IC	Intelligence Community
ICAM	Information Collection for Automated Mapping
IJIS	Integrated Justice Information Systems
ILP	Intelligence-Led Policing
IRS	Internal Revenue Service
IT	Information Technology
JXDM	Justice (Extensible Markup Language) Data Model
LE	Law Enforcement
LEAA	Law Enforcement Alliance of America
LEADS	Law Enforcement Agency Data System
LEIN	Law Enforcement Intelligence Network
LEIU	Law Enforcement Intelligence Unit
LEO	Law Enforcement Online
MO	Modus Operandi
NCISP	National Criminal Intelligence Sharing Plan
NDIC	National Drug Intelligence Center

NIBRS	National Incident-Based Reporting System
NIM	National Intelligence Model
NLECTC	National Law Enforcement and Corrections Technology Center
NSA	National Sheriffs' Association
NTL	National Training Laboratory
NW3C	National White Collar Crime Center
OJP	Office of Justice Programs
OLAP	Online Analytic Processing
PD	Police Department
POP	Problem-Oriented Policing
PSNI	Police Service of Northern Ireland
RCMP	Royal Canadian Mounted Police
RICO	Racketeer Influenced and Corrupt Organizations Act
RISS	Regional Information Sharing Systems
SARA	Scanning, Analysis, Response, Assessment
SPSS	Statistical Package for the Social Sciences
UCR	Uniform Crime Reporting
ViCap	Violent Criminal Apprehension Program
XML	Extensible Markup Language

Chapter 1

UNTAPPED CAPACITIES TO
ANALYZE INTELLIGENCE

Since 9/11, national security agencies and law enforcement agencies are seeking to build unprecedented partnerships. The urgent need to identify and prevent potentially destructive actions by those who threaten to harm us as a nation on our own territory demands new alliances. The challenge of combining the "eyes and ears" on our streets, or local level law enforcement, with the resources of federal law enforcement agencies and national security entities is great. The London suicide bombings in July 2005 emphasize the need for local-level knowledge to address future threats of terrorism. New ways of thinking to achieve a more secure homeland are not only desirable, but also essential to our continued survival.

This book explores analytical capabilities in law enforcement, with a focus on local applications. Along with those in the political and media arenas, the 9/11 Commission has not recognized that intelligence analytical capacities exist in state and local law enforcement, and little mention of this emerging resource exists in the literature of the war on terrorism, or the Long War.

The purpose of this book is to inform the larger community of federal government agencies, including law enforcement, national security, and other interested entities, as well as the citizens of this country and beyond, about the intelligence analytical capabilities existing in local and state levels of law enforcement.

This work challenges the thinking of the national Intelligence Community and its analysts, as well as the law enforcement community, by using an organizational change management process called "Appreciative Inquiry." Appreciative Inquiry focuses on using imagination, the very thing found lacking in the U.S. Intelligence Community in evaluations of intelligence failures. The first stage of this process, the "discovery stage," is incorporated into this work through success stories revealed in the author's interviews with analysts and experts who have contributed to real-world analytical work in law enforcement. Those success stories illustrate local law enforcement analytical capabilities.

Untapped Information

Type of agency	Number of agencies	Number of full-time sworn officers
Total		796,518
All State and local	17,784	708,022
Local police	12,666	440,920
Sheriff	3,070	164,711
Primary State	49	56,348
Special jurisdiction	1,376	43,413
Texas constable	623	2,630
Federal*		88,496

Note: Special jurisdiction category includes both state-level and local-level agencies. Consolidated police-sheriffs are included under local police category. Agency counts exclude those operating on a part-time basis.
*Non-military federal officers authorized to carry firearms and make arrests.

In 2000, nearly 800,000 full-time, sworn law enforcement officers worked in the U.S.

Source: http://www.ojp.usdoj.gov/bjs/lawenf.htm.

State and local law enforcement officers, given their number, collect an enormous amount of information. Untold, untapped quantities of information exist within their reports that, if analyzed, could help solve many crimes, help direct efforts to prevent crime, and, possibly, help find and/or connect some of the dots to help prevent terrorist acts in the United States.

The collectors are producing — but who is available to analyze the information? The detailed reports of state and local law enforcers stay within their respective agencies. Federal agencies do not obtain the data contained in such reports. State law enforcement agencies are likely to have analysts to support investigations, but many local law enforcement agencies have no personnel dedicated to information analysis. Analysts do exist at the local level to carry out this task and some of their work is excellent — but an understanding of their role and support to develop their profession is absent.

Former CIA intelligence methodologist Richards J. Heuer noted, "Major intelligence failures are usually caused by failures of analysis, not failures of collection."[1] It is hardly a stretch of the imagination to consider that crime prevention and criminal apprehension efforts fall short in part because of lack of analysis in law enforcement.

---

[1] Richards J. Heuer, Jr. *Psychology of Intelligence Analysis* (Washington, DC: Center for the Study of Intelligence, 1999), Chapter Six. URL *http://www.cia.gov/csi/books/19104/art9.html.*

With the effective application of law enforcement intelligence analysis, terrorists resident in the U.S. who have had documented contact with terrorist networks can be apprehended by state and local level law enforcement. For example, on 9 September 2001, Ziad Jarrah, one of the 9/11 hijackers, was stopped by a veteran Maryland state trooper who did not know Jarrah was an individual of concern to the U.S intelligence community.[2] Similarly, a command officer from the Washington State Patrol believes, in retrospect, that he unwittingly saw terrorist Mohammed Atta check out of a motel in Portland, Maine on the morning of 11 September 2001.[3] Non-federal law enforcement is much more likely to see potential terrorists as they go on with their day-to-day business. State- and local-level analysts are likely to have data related to such contacts in the form of field interview reports or 911 calls that might, on the surface, appear innocuous. With unimpeded access to federal databases, local law enforcement intelligence analysts can support offices in the field with the available, but often missing, linkages for effective law enforcement.

Conversely, detailed data on calls for service and crime data maintained in police departments belong to the individual agency and are typically not accessible to the FBI and other law enforcement agencies. This means that when an offender commits a crime or if there is unusual activity in one jurisdiction, it is not possible for anyone to correlate it with another jurisdiction, unless there is regional data sharing and/or analysts working to share this type of information. Regional data sharing is emerging in some parts of the country, but without proactive analysis of the data, series and patterns will not be discovered. For the most part, law enforcement analysis is applied to current investigations and identified crime problems, the things we already know about. Anticipatory discovery, or the application of intelligence principles to law enforcement analysis, only occurs in agencies where analysts have the freedom and the ability to initiate inquiries and to recognize new problems.

Although most law enforcement agencies report crime statistics to the FBI, the data they send are aggregated and do not give specifics. Even National Incident-Based Reporting System (NIBRS) data,[4] with more detail than the FBI's Uniform Crime Report,[5] provide little qualitative information of sufficient specificity for effective analysis. Good qualitative data provide the specifics needed to identify serial criminals or suspicious activities or anomalies. Few serial crime patterns or serial criminals can be identified in the absence of such data. In practice, no terrorist can be identified without sound and extensive qualitative data.

Statistical information, as indicated in the pyramid above, is not scarce. However, as one analyst told the author, "The 1980s was supposed to be the Information Age, yet,

---

[2] Jonathan R. White, *Defending the Homeland* (Belmont, CA: Wadsworth, 2004), 2.

[3] White, 3.

[4] For a description of NIBRS data see *http://webapp.icpsr.umich.edu/cocoon/NACJD-STUDY/03449.xml.*

[5] See *Summary of Uniform Crime Reporting (UCR) Program* at *http://www.fbi.gov/ucr/cius_00/00crime1.pdf.*

although we collected tons of data we did not analyze it. We were constantly collecting data to count crimes but we did not have systems set up to search information to solve crimes." This trend continues in many law enforcement agencies to this day.

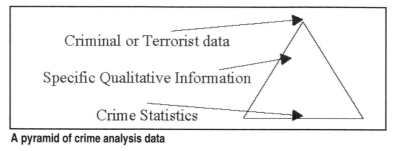

A pyramid of crime analysis data
Source: Author

The 9/11 Commission noted that, in looking at individual cases, no one looked at the bigger picture to see the blinking lights prior to the 11 September terrorist attacks.[6] Likewise, no one proactively looks now at the bigger picture by analyzing the qualitative data from law enforcement to look for connections. Many data already exist and have been collected for decades, but they are collected for the purpose of counting. The data are not used to identify on a broad scale, for example, chronic low-level crime offenders, situations that foster crime, or changing modus operandi that might be indicators of organized crime. Since we can expect that terrorism is often funded by criminal activities, doing all possible to examine the information we already possess becomes even more important than its obvious usefulness in combating and potentially reducing the domestic crime that plagues our nation.

The FBI employs over 28,000[7] — but its staffing is the equivalent of only five per cent of local-level law enforcement numbers. How do we maximize use of our numerous national law enforcement personnel? It is time to tap the resources we have through exploring the possibilities of analysis at the state and local level of law enforcement. Developing the role of the analyst of criminal activity can be a significant force multiplier in improving public safety, both for crime reduction and for homeland security.

History

Law enforcement in the United States, unlike the military, has not been intelligence-driven, until recently, and even now most often is not. Although crime analysis and intelligence analysis have existed in various state and local law enforcement agencies since the 1970s, intelligence processes are not institutionalized nor widely understood by law enforcement managers and officers.

---

[6] *9/11 Commission Report, (Washington, D.C.: U.S. Government Printing Office, 2004), 254.*
[7] *http://www.fbi.gov/fbihistory.htm.*

Intelligence analysis developed in law enforcement in the 1940s and 1950s to tackle organized crime, and in the 1960s intelligence units appeared in some large local law enforcement agencies.[8] Passage of the Racketeer Influenced and Corrupt Organizations (RICO) legislation in 1970 influenced the development of intelligence analysis in law enforcement.[9] RICO's focus on organized crime has influenced the development of what is commonly referred to as "intelligence analysis" in law enforcement. This type of analysis tends to look for networks of criminal activity with the ultimate goal of prosecuting high-level operators in organized crime networks.

The term "crime analysis" was first offered in the second edition (1963) of Orlando Winfield Wilson's book, *Police Administration*:[10]

> **Crime Analysis.** The crime-analysis section studies daily reports of serious crimes in order to determine the location, time, special characteristics, similarities to other criminal attacks, and various significant facts that might help to identify either a criminal or a pattern of criminal activity. Such information is helpful in planning the operations of a division or district.

Funding from the Law Enforcement Assistance Administration in the 1970s also contributed to the development of crime analysis; crime analysis was a key pillar of the Law Enforcement Alliance of America (LEAA's) Integrated Criminal Apprehension Program.[11]

---

[8] *Intelligence 2000: Revising the Basic Elements,* Marilyn B. Peterson, ed., (n.p.: a joint publication of the Law Enforcement Intelligence Unit (L.E.I.U.) and International Association of Law Enforcement Intelligence Analysts (IALIA), 2000), 3-4.

[9] *Intelligence 2000: Revising the Basic Elements,* 4.

[10] International Association of Crime Analysts, *Exploring Crime Analysis* (Overland Park, KS: IACA Press, 2004), 14.

[11] *Exploring Crime Analysis,* 14.

The Two Pathways to Analysis

Thus, in the United Sates, law enforcement intelligence analysis at the state and local levels has developed on two distinct pathways. The first pathway is more familiar to those at the federal level. It consists of the intelligence analysis traditions that exist in the FBI and most other federal law enforcement agencies, namely, to support investigative operations. The other pathway is that found mostly at the local level of law enforcement— crime analysis. Crime analysis in local level-law enforcement assists decisionmakers tactically and strategically. It exists at the state level to study larger crime trends and measure changes in crime levels. Crime analysis is similar to some aspects of national-security analytical work. The crime analyst may look at a whole city to see where problems are, just as a national security analyst may look at a whole country to try to understand emerging or existing problems that might affect U.S. security interests.

Marilyn B. Peterson, past president of the International Association of Law Enforcement Intelligence Analysts, writes in *Applications in Criminal Analysis:*
Intelligence or investigative analysis has generally been held separate from "crime analysis," which, in its simplest form, is done within police departments to determine patterns of burglaries, auto thefts, and so on. Crime analysis has been a patrol-oriented form of analysis: that is, its conclusions are often used to support decision making on the deployment of patrol officers. That deployment is often done as a measure to attempt to prevent further crimes of the type that was the subject of the crime analysis. Investigative [intelligence] analysis, on the other hand, has been used to support the major crimes or organized crime functions within law enforcement, to help solve a particular case.[12]

As one subject interviewed for the present project put it, "Crime analysts are much more involved in analyzing crimes, and intelligence analysts are much more involved in analyzing criminals."

The distinctions between these types of analysis are peculiar to the structure of law enforcement in the U.S. In this study, in professional debates and in the literature, law enforcement analysts from other countries state that they do not understand why two tracks have developed as separate approaches in the U.S. There are a number of possible reasons for this development. However, the main reason is likely to be the number of law enforcement agencies in the United States and the "silo" effect of having so many separate agencies with different missions specializing in particular crimes and investigations.

The type of analysis done in law enforcement is related to the mission and needs of the agency. This separation of analytical roles between crime analysts and investigative analysts is similar to the segregation or specialization among agencies in different locations and at differing levels, which leads to withholding information, and to the enduring sense that each agency takes a unique approach to work. Virtually all of the skills used in crime analysis can

---

[12] Marilyn B. Peterson, *Applications in Criminal Analysis: A Sourcebook* (n.p.: Praeger/Greenwood, 1998), 2.

be applied to intelligence analysis. The intelligence analyst may have to learn new ways of looking at crime in order to cross over to the crime analysis field, but the critical thinking skills used to analyze organized crime and support investigators are applicable as well to the analysis of street crimes. The technical requirements of criminal analysis and intelligence analysis vary, but in any case, specific software and analytical products also vary from agency to agency. The requirements that analysts have the ability to learn software applications and create customer-based analytical products are the same for both positions.[13]

In November 2004 the Global Justice Information Sharing Initiative and the International Association of Law Enforcement Intelligence Analysts, Inc. (IALEIA), through funding provided by the U.S. Department of Justice's OJP (Office of Justice Programs), published the booklet *Law Enforcement Analytical Standards.*[14] This publication points out the resource-conserving role of intelligence analysis:

> In general, the role of law enforcement intelligence is to help agencies reduce crime patterns and trends. The intended result of law enforcement is to lower crime rates, whether through apprehension, suppression, deterrence, or reduced opportunity. Analysis supports good resource management and is directly involved in creating situational awareness, in assisting decision making, and providing knowledge bases for law enforcement action.[15]

In March 2005, a focus group consisting of experts in crime analysis convened in Ashburn, VA, tasked with developing standard functional specifications for law enforcement records management systems. The host group, the Law Enforcement Information Technology Standards Council (LEITSC) comprises representatives from the International Association of Chiefs of Police, the National Sheriffs' Association, Police Executive Research Forum and the National Organization of Black Law Enforcement Executives. The focus group recommended that the term *crime analysis* be changed to *analytical support* and that analysts be given the authority to access and export all data in such systems for whatever types of analyses needed. Thus, specifications for records management systems would reflect the recognition that all types of information analyses may be potentially useful in supporting the missions of law enforcement.[16]

Although some law enforcement analysts are sworn officers or agents, the trend is to deploy civilians as analysts. Due to the increasing need for advanced skill levels in analytical work, and the aspect of sworn officers' work that allows them to frequently transfer positions, using

---

[13] Also see California Occupational Guide 557 (1999) with the title "Crime and Intelligence Analysts" at *http://www.calmis.cahwnet.gov/file/occguide/CRIMANLT.HTM.* California was the first state to set standards for crime and intelligence analysts in law enforcement. Although the definitions may not be up-to-date nor comprehensive, they do differentiate between the two roles of the analyst in law enforcement.

[14] Available at *http://it.ojp.gov/documents/law_enforcement_analytic_standards.pdf.*

[15] Please see *http://it.ojp.gov/documents/law_enforcement_analytic_standards.pdf,* 3.

[16] Meeting of the Law Enforcement Information Technology Standards Council focus group on crime analysis standards for RMS in Ashburn, VA, March 23-25, 2005.

civilians as analysts helps establish a consistent, experienced cadre of professional workers. Training individuals who will stay in this field is cost-effective. Improving analysts' wages and developing career paths will encourage growth of the profession.

Civilian analysts free up sworn personnel to patrol, investigate, and do the work needed in the field. This does not mean that sworn personnel should not know how to analyze information; rather, it means that dedicated personnel should be consistently deployed in the increasingly sophisticated and specialized work of law enforcement analysis.

It is very important that our first line of defense against terrorism — the seven hundred thousand officers on the street — be given adequate training and background information on terrorism, the methods and techniques of the terrorists, and the likelihood of an imminent attack. The reason this information should be shared, or available, is not so that state and local police can be involved in the investigation of terrorist cells, or individual terrorists, or collecting raw intelligence information. Rather it is simply that these officers know their territory and are on the street 24 hours a day. Considering that the terrorists who attacked the World Trade Center were stopped on several occasions by the local police prior to the attack for minor traffic violations, it is logical to assume that this pattern of random interception would continue in the future. If and when similar situations occur, our local and state officers should have background knowledge by which to arrive at a reasonable suspicion. Thereafter, the officers should have the ability to access national data banks to assist in the further resolution of the matter at hand. This is a largely ignored but critical asset in our struggle to contain terrorism.[17]

Let us consider a missing element in the scenario above — trained and educated people to analyze the information collected by the 700,000 officers.

Inadequacies of Law Enforcement Analysts Federally and Locally

Prior to 9/11, federal law enforcement agencies were moving toward legitimizing the role of the analyst. In 2000, the FBI was in the process of developing competencies for analysts in order to ensure a minimal level of effectiveness and quality in performance.[18] The DEA, as part of its counterdrug initiatives, had developed new training modules for analysts and was creating a career track similar to that of special agents in order to retain quality staff.[19] Subsequently, the 9/11 Commission identified common, serious problems in the field of law enforcement intelligence analysis: lack of proactive analysis, hiring unqualified persons, and inadequate information technology.

---

[17] Major Cities Chiefs Association Intelligence Commanders Conference Report, "Terrorism: The Role of Local and State Police Agencies" in *Terrorism: The Impact on State and Local Law Enforcement*, (June 2002), 5. http://www.neiassociates.org/mccintelligencereport.pdf.

[18] Gary Burton, Presentation on FBI Analyst Competencies on 7 June 2000 at Mercyhurst College, Erie, PA.

[19] Jill Webb, Presentation on DEA Analysts Training on 7 June 2000 at Mercyhurst College, Erie, PA.

However, the FBI had little appreciation for the role of analysis. Analysts continued to be used primarily in a tactical fashion—providing support for existing cases. Compounding the problem was the FBI's tradition of hiring analysts from within instead of recruiting individuals with the relevant educational background and expertise. Moreover, analysts had difficulty getting access to the FBI and intelligence community information they were expected to analyze. The poor state of the FBI's information systems meant that such access depended in large part on an analyst's personal relationships with individuals in the operational units or squads where the information resided. For all of these reasons, prior to 9/11 relatively few strategic analytic reports about counterterrorism had been completed. Indeed, the FBI had never completed an assessment of the overall terrorist threat to the U.S. homeland.[20]

Systematic information analysis itself in *all* law enforcement is in its infancy. It is seldom proactive. Intelligence analysis in most federal agencies revolves around existing investigations rather than uncovering new threats. This is changing in the post-9/11 world, but a great obstacle to change is a lack of skilled staff to analyze information and systems that support the growth of intelligent *intelligence* analysis.

Captain William S. Brei touches on this concept in *Getting Intelligence Right: The Power of Logical Procedure.*

Although all intelligence activities are conducted to answer documented requirements, analysts who understand a customer's circumstances may find it possible to increase the product's relevancy by answering unstated — or unforeseen — questions that present themselves in the data and in the analysis.[21]

Looking for the unknown threats that are hidden in the data requires the proactive and effective analysis of data — people with a proclivity to answer the *unstated* and *unforeseen* questions. There may be unknown elements existing in the data — hidden threats that we could uncover, if we had adequate resources dedicated to the task.

Developing systematic crime and intelligence analysis at the local level of law enforcement remains a challenge due to the sheer number of local law enforcement agencies in our nation. Each agency is separate and has a mission to protect and serve within its own boundaries. Each agency is beholden to its taxpayers to make their concerns the highest priority — priorities that often exclude information sharing across jurisdictions, as well as information sharing for federal priorities. Many local law enforcement agencies are very small and do not have the resources nor crime volume to justify development of their own analytical units, nor the urgency to share their information with a larger pool of governmental agencies.

---

[20] *9-11 Commission Report*, 77.

[21] Captain William S. Brei (USAF), *Getting Intelligence Right: The Power of Logical Procedure,* Occasional Paper Number Two (Washington, DC: Joint Military Intelligence College, January 1996), 22.

The March 2003 report *Crime Analysis in America: Findings and Recommendations*[22] by primary investigator Timothy O'Shea, funded by the Office of Community-Oriented Policing Services (COPS), U.S. Department of Justice, describes the vague work situation of the crime analyst in America:

> We found that only one of the departments that we visited or spoke with had a formal job manual for the crime analyst position. Several were in the process of drafting one. Nearly every analyst that we spoke with believed that a manual was necessary, but they also pointed out that drafting one was a complex and difficult task that required a collaborative effort between analysts and managers. The general absence of a formal manual further illustrates the ad hoc nature of crime analysis in American law enforcement and is a further indication that the function has not been given the careful, deliberate consideration that it should.[23]

The Challenge of Conceptualizing "Intelligence" in Law Enforcement

In the National Intelligence Community, "intelligence" refers to information affecting national security decisionmaking. In the military, "intelligence" often refers to information that is helpful in understanding how to combat enemy forces. "Intelligence" in law enforcement generally has referred to information gathered from police officers or law enforcement agents noting suspicious activities, information provided by police informants, and information gleaned by investigators through investigations. It is not necessarily obtained secretly or covertly, although it may be. Of course, in all environments, intelligence is more than mere information.

> If you don't know the difference between information and intelligence then you don't see the value of analysis.[24]

> More than 40 years after Sherman Kent, the CIA's father of intelligence analysis, persuasively argued that intelligence is knowledge, some still confuse the method with the product. Sadly, such confusion is widespread.[25]

---

[22] See *http://www.cops.usdoj.gov/mime/open.pdf?Item=855*.

[23] See *http://www.cops.usdoj.gov/mime/open.pdf?Item=855,17*.

[24] Quote from author's interview with a subject who prefers to remain anonymous.

[25] Stephen C. Mercado, "Sailing the Sea of OSINT in the Information Age," *Studies in Intelligence* 48, no. 3 (2004), 51 at *http://www.cia.gov/csi/studies/vol48no3/article05.html*.

Intelligence analytical products in law enforcement are usually considered "intelligence" if they are about specific individuals or groups. Tactical crime analysis products, a crime series bulletin for example, are generally not called "intelligence" in local law enforcement agencies. There is no consensus regarding how and when to call information *intelligence*. Some experts say unprocessed information differs from true "intelligence," and that *intelligence* is the value added by *analysis* and *interpretation* within specific contexts.

New York City's Police Department has instituted a process called COMPSTAT[26] that is dependent on four components, the first being timely and accurate intelligence. In this context, "intelligence" means all information that might be useful, whether raw data or information processed by analysis. Perhaps, to the recipient, it is the utility of information that defines it as intelligence. Thus, all information that supports decisionmaking in the intelligence cycle, whether it be simply a wiretap providing a crucial name or a comprehensive study of human smuggling across international borders with detailed recommendations to prevent this activity, would be "intelligence."

Diluting Intelligence by Calling Everything Intelligence

Clarity in understanding the meaning of intelligence is needed if the role of the analyst is to remain healthy and to grow. This applies to national security as well as to law enforcement. In a society in which information access and production are growing exponentially, Wilhelm Agrell's concern that intelligence be rigorously defined applies:

> The problem is not in law enforcement but in the application of intelligence analysis in fields where its specific virtues are not adequate, not actually needed, or even might become counter-productive. Intelligence analysis can, employed in the right context, considerably enhance over-all performance. But in the wrong context "intelligence" could be just another dead weight — wasting resources, complicating procedures, or creating unrealistic expectations of gains or results. What I am referring to could be described as the application of the concept or perhaps the illusion of intelligence analysis to various information-processing activities that are not really intelligence in the professional sense of the word. Broadening the concept is one thing — to flatten it out is something quite different.[27]

This research project, by virtue of interviewing and capturing the values of experienced analysts and related experts, aims to expand the concept of intelligence in law enforcement and add to its understanding in the larger public safety and security community.

---

[26] "CompStat (short for COMPuter STATistics or COMParitive STATistics) is the name given to the New York City Police Department's management accountability process. CompStat is a multi layered, dynamic approach to crime reduction, quality of life improvement and, personnel and resource management."

[27] Wilhelm Agrell, University of Lund, Sweden, *When everything is intelligence — nothing is intelligence*, The Sherman Kent Center for Intelligence Analysis Occasional Papers: 1, No. 4 (October 2002), 3.

Defining Intelligence and Intelligence Analysis in the IC

Even though it has a longer history, defining intelligence analysis in the Intelligence Community is also a challenge, according to Stephen Gale:

> Just what is this intelligence business anyway? What is it that these "intelligence professionals" (the "intelligence communities" of the world) do to earn their keep? What are their professional standards, their backgrounds, training, risks and rewards? And, short of having worked in the intelligence field, how can we, as engaged citizens, decide whether to be shocked or awed by accounts that point to failures in political decisions that are based on intelligence analysis?

> Hard and fast rules are difficult to find in the world of intelligence analysis. More often than not, intelligence analysts are faced with highly complex circumstances, imperfect information, obscure motivations, conflicting interpretations, and deep ideological and political differences. Simple answers are not only in short supply but, when invoked, have the downside consequence of being misleading or worse.[28]

The crime and intelligence analyst in law enforcement is at an advantage in that the world of law enforcement is information-rich, even though some of the information is inaccessible due to technological failures, lack of computer-literate staff, jurisdictional rivalries, and plain old lack of understanding by management (and sometimes analysts) as to what tools and information are needed to succeed.

Intelligence: The Reality of Context

It may be helpful, in the course of dialogue, discussion, and debate regarding "intelligence analysis," to consider that the arguments and positions we have and take are limited by our own experience and areas of responsibility.

> In the course of argument, people frequently complain about words meaning different things to different people. Instead of complaining, they should accept it as a matter of course. It would be startling indeed if the word "justice," for example, were to have the same meaning to each of the nine justices of the United States Supreme Court; then we should get nothing but unanimous decisions. It would be even more startling if "justice" meant the same thing to the robber as to the robbed. If we can get deeply into our consciousness the principle that no word ever has the same meaning twice, we will develop the habit of automatically examining contexts, and this enables us to understand better what others are saying. As it is, however, we are all too likely to have automatic, or signal, reactions to certain words and read into people's remarks meanings that were never intended. Then we waste energy in angrily accusing people of intellectual dishonesty or abuse of words,

---

[28] Stephen Gale, *Standards of Intelligence Reasoning*, in Foreign Policy Research Institute: A Catalyst for Ideas,14 November 2004, http://www.fpri.org.

when their only sin is that they use words in ways unlike our own, as they can hardly help doing, especially if their background has been widely different from ours.[29]

Thus, intelligence has its context for each of us. To understand how our views vary yet reach a consensus is the challenge. In the judgment of the present author, after study and exposure to analysts in various venues, intelligence analysis is not fundamentally different in the military, national security, or law enforcement. Unless the reader has had similar exposure, he or she may disagree. Critical thinking requires that one suspend judgment until all the available facts are presented and their applicability to alternative definitions is ascertained. Perhaps, at the end of this book, those readers who disagree now will have changed their minds.

The New Emphasis on Intelligence in Law Enforcement

Although the core concept of intelligence analysis may not be different within the various entities that conduct it, law enforcement has only recently begun to emphasize its utility. This emphasis most notably began with the unveiling of the United Kingdom's "National Intelligence Model" in 2000.[30]

This work is the outcome of a desire to professionalize the intelligence discipline within law enforcement. Intelligence has lagged behind investigation in the codification of best practices, professional knowledge and in the identification of selection and training requirements of staff. It is also recognition of the changing requirements of law enforcement managers which highlights three particular needs:

- To plan and work in co-operation with partners to secure community safety
- To manage performance and risk
- To account for budgets[31]

The model addresses three levels: local, cross-border (which refers to multiple agencies working together), as well as serious and organized crime.[32] It focuses on intelligence as a business with measurable outcomes.

Similar developments in the United States are more recent. In August 2002, The International Association of Chiefs of Police worked with other entities to publish *Criminal*

---

[29] S.I. Hayakawa and Alan R. Hayakawa, *Language in Thought and Action* (Orlando, FL: Harcourt, Inc., 1990), 40.

[30] The National Intelligence Model: Providing a Model for Policing. URL *http://www.policereform.gov.uk/implementation/natintellmodeldocument.html.*

[31] The National Intelligence Model: Providing a Model for Policing, 7.

[32] The National Intelligence Model: Providing a Model for Policing, 8.

Intelligence Sharing: A National Plan for Intelligence-Led Policing.[33] The concept of intelligence-led policing as a medium for agency integration is encapsulated in the diagram below. Later, the Office of Community-Oriented Policing contracted research resulting in the November 2004 publication of *Law Enforcement Intelligence: A Guide for State, Local, and Tribal Law Enforcement Agencies.*[34] The concept of Intelligence-Led Policing is discussed later in the present work.

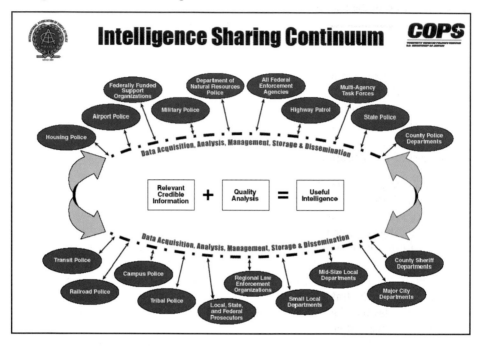

Source: URL *http://www.theiacp.org/research/CriminalIntelligenceSharingReport.pdf,* between pages 4 and 5.

---

[33] U.S. Department of Justice (DOJ), Community-Oriented Policing Services, and International Association of Chiefs of Police, *Criminal Intelligence Sharing: A National Plan for Intelligence Led Policing* (n.p., 2002). URL *http://www.theiacp.org/research/CriminalIntelligenceSharingReport.pdf.*

[34] U.S. DOJ, Community-oriented Policing Services (COPS), *Law Enforcement Intelligence: A Guide for State, Local, and Tribal Law Enforcement Agencis,* published at *http://www.cops.usdoj.gov/default.asp?Item=1404.*

The Criminal Intelligence Coordinating Council

The Criminal Intelligence Coordinating Council (CICC) was established in May 2004. It is in the very early stages of providing direction for state and local law enforcement to improve understanding of information and intelligence collection, analysis, and sharing. Establishment of the CICC reflected the complex interaction of local and national concerns as illustrated in the text box below.

Bold Beginnings[35]

In order to address past inadequacies in our nation's intelligence process, the U.S. Department of Justice's Office of Justice Programs (OJP), at the request of the International Association of Chiefs of Police (IACP), authorized the formation of the Global Justice Information Sharing Initiative (Global) Intelligence Working Group (GIWG), one of several issue-focused subgroups of the Global Advisory Committee, to develop a collaborative intelligence sharing plan.

The GIWG supported the development of the National Criminal Intelligence Sharing Plan (NCISP) as a blueprint to assist law enforcement personnel in their crime-fighting, public safety, and counterterrorism efforts. The plan acknowledged that officers, investigators, and analysts working throughout the United States are the first line of prevention and defense against terrorism and crime. The NCISP recognized the importance of local, state, and tribal law enforcement agencies as a key ingredient in the nation's intelligence process and called for the creation of the CICC to establish the linkage needed to improve intelligence and information sharing among all levels of government. Made up of members from law enforcement agencies at all levels of government, the CICC was created to provide advice in connection with the implementation and refinement of the NCISP. Members of the CICC serve as advocates for local law enforcement and support their efforts to develop and share criminal intelligence for the purpose of promoting public safety and securing our nation. These goals are attainable and necessary for the continued safety of U.S. citizens and visitors.

Developing and sharing criminal intelligence will be a challenge in the United States due to the fragmented, many-layered structure of its law enforcement system.

Why Change is Mandated

Terrorists commit "crimes ranging from: immigration violations, sale of contraband cigarettes, money-laundering, mail fraud, credit card fraud, insurance fraud, racketeering and

---

[35] *Police Chief Magazine* 72, no. 2 (February 2005). URL *http://policechiefmagazine.org/magazine/index.cfm?fuseaction=print_display&article_id=512&issue_id=22005.*

arson."[36] If we are to be effective in preventing terrorism, analyzing all crime is necessary. The linkage of crime, even street crime, to terrorism is widely documented.

Many transnational terrorist groups engage in low-level criminal activity to fund operations, facilitate movement of personnel and procure weapons and explosives. Their activities include robbery, credit card fraud, passport and identity forgery, drug trafficking and immigration violations.

Worldwide, much of this activity centres on localised groups of disaffected young men, often from diaspora communities, who become ensnared in militancy via their criminal activities or while in prison.[37]

The emerging changes in society also call us to change. Individuals commit crimes and individuals commit acts of terror.

The power of the individual: new technology will allow ever smaller, more anonymous groups or individuals to commit crimes previously beyond their means. We are entering the age of asymmetric attacks, where individuals can take advantage of their relative anonymity to strike quickly and without trace against targets at both a national and international level. The more systems are networked, the greater the danger from a single attack. New technology allows people to group together more easily, overcoming geographical limitations. There is also the concern that organised crime will steer or fund these individuals into pursuing its goals.[38]

The growing need to understand the nature of low-level crime in relationship to terrorist activities calls us to think and act in new ways. The growing need to create a new network that acts intelligently and proactively to interfere in and disrupt the activities of criminal and extremist groups is obvious.

Disrupting Terrorists: A Systems Approach

It is time we use our law enforcement resources in a new manner. In the last fifteen years, new law enforcement strategies, with a foundation in analysis of information, have emerged to change the focus in law enforcement work to prevention of crime rather than the traditional focus on apprehension and prosecution, which often do not result in a reduction of crime. Crime, like terrorism, can be prevented. The strategies now being developed to prevent crime may be adapted to prevent terrorism and vice versa. New concepts will emerge as we learn to think of the problem in a broad, system-based manner, possibly in terms of the concept of "safety." For, after all, whether a military mission, a

---

[36] Christopher Jasparro, "Low-level Criminality linked to Transnational Terrorism," *Jane's Intelligence Review* 17, no. 5 (May 2005), 19.

[37] Jasparro, "Low-level criminality Linked to transnational terrorism," 18.

[38] UK Crime Prevention Panel, Department of Trade and Industry, *Turning the Corner* (December 2000), 11 URL *http://www.foresight.gov.uk/Previous_Rounds/Foresight_1999_2002/ Crime_Prevention/Reports/Summary_of_Action/index.htm.*

national security mission, or a law enforcement mission, the bottom line is this: the mission is safety. Physical, economic, political, social and personal safety lie at the core. Our systems to ensure the safety of citizens are often working at cross-purposes because the overarching mission is lost in the course of our giving attention to the differences in the ways we work and think. This study expects to bridge some of those differences.

At times, law enforcement and intelligence have competing interests. The former head of the FBI's international terrorism division notes that Attorney General Reno leaned toward closing down FISA surveillance if they hindered criminal cases. White, however, notes that the need for intelligence was balanced with the effort to arrest and prosecute terrorists. In addition, as noted earlier, convictions that help disrupt terrorists are often on minor charges (such as immigration violations), which do not always convince Field Office personnel that the effort is worthwhile compared with putting criminals in jail for many years. As former FBI Executive Assistant Director for Counterterrorism and Counterintelligence Dale Watson explains, Special Agents in Charge of FBI Field Offices focused more on convicting than on disrupting.[39]

The Bottom Line

The U.S. Homeland Security Strategy expects law enforcement professionals to exhibit a capacity to analyze information. The present study builds on the four assumptions and preliminary findings listed below to address the issue of how law enforcement professionals in the U.S. are adapting and adopting tools of analysis and synthesis to improve personal and homeland security.

- Law enforcement crime and intelligence analysis play important roles in maintaining and improving public safety and homeland security.
- Crime and intelligence analysts at every level can learn and adapt best practices from one another and collectively.
- The roles of law enforcement crime and intelligence analysts are not widely understood nor documented — this study aims to improve upon this situation.
- In some cases, law enforcement crime and intelligence analysts operate with fewer restrictions and bureaucratic obstacles than those in federal agencies; thus in most cases, analysts at the local level have more freedom to be creative and develop innovative analytical techniques/approaches that may help the wider public safety/national security community.

This book builds on law enforcement crime and intelligence analysis — combining material gleaned from what is already written about the subject, and information gathered through interviews with working analysts and experts who have directly contributed to law enforcement analysis as a field. The next chapter explains the research methods and the concept of Appreciative Inquiry as a process to engage imagination and accelerate positive change in the field.

---

[39] Eleanor Hill, Staff Director, Joint Inquiry Staff, Joint Inquiry Staff Statement: *Hearing on the Intelligence Community's Response to Past Terrorist Attacks against the United States from February 1993 to September 2000*, 8 October 2002, 18.

Chapter 2

RESEARCH THROUGH APPRECIATIVE INQUIRY

Appreciative Inquiry (AI) is a paradigm for organizational development through "action research," conceptualized in 1980 by Dr. David Cooperider and Suresh Srivastva at Case Western Reserve University in Cleveland, Ohio. AI has been used by the U.S. Navy to develop leadership skills, by the Roadway Express Company to study best practices and accelerate change, and by the city of Chicago in a project called "Vision Chicago." Examples of other companies/entities that have used AI include McDonald's, John Deere, GTE, Texas Insurance, the Catholic Church, Habitat for Humanity, the United Nations, and the Environmental Protection Agency.[40]

In the comprehensive *Handbook for Qualitative Research*, AI is grouped with emerging research methods representing significant paradigm shifts from traditional research methods. Appreciative inquiry as described in that text is a *relational process*, rather than a research methodology, one that "wholly transforms relationships among otherwise hostile members of an organization or community."[41] A medium of human communication familiar to all those who inhabit organizations, whether large or small, is storytelling. Storytelling is a core element of the relational process that binds humans together. For this research project, appreciative inquiry was extended to the law enforcement workplace by asking interview subjects to tell stories that relate their work experiences to the development, acceptance and use of concepts and tools that accompany the professional evolution of the field of crime analysis, or law enforcement intelligence. The appreciative inquiry approach to organizational change and improvement allowed the author to capture a sense of the strengths, successes, values, hopes and dreams of individuals who are charged with promoting homeland security through law enforcement activities.

AI is used in this research to challenge the mindset of analysts and readers, in response to retired CIA "philosopher" Heuer's call for intelligence analysts in the Intelligence Community to explore *different mental models*.

> Today, there is greatly increased understanding that intelligence analysts do not approach their tasks with empty minds. They start with a set of assumptions about how events normally transpire in the area for which they are responsible. Although this changed view is becoming conventional wisdom, the Intelligence Community has only begun to scratch the surface of its implications.

---

[40]See *Appreciative Inquiry Commons* at http://appreciativeinquiry.cwru.edu/. Also see Jane Magruder Watkins & Bernard Mohr, *Appreciative Inquiry: Change at the Speed of Imagination* (San Francisco, CA: Jossey-Bass/Pfeiffer, 2001).

[41]Norman K. Denzin, and Yvonna S. Lincoln, eds., *Handbook of Qualitative Research*, 2d ed. (Thousand Oaks, CA: Sage Publications, 2000), 1039.

If analysts' understanding of events is greatly influenced by the mind-set or mental model through which they perceive those events, should there not be more research to explore and document the impact of different mental models? [42]

AI is designed to influence mental models and change processes by radically altering the way we normally perceive and "solve" problems. The process seeks to uncover what is already working in a system and allows participants to hear what others value in the system. It has shown itself effective as a change agent across many disciplines.

Appreciative Inquiry is about the coevolutionary search for the best in people, their organizations, and the relevant world around them. In its broadest focus, it involves systematic discovery of what gives "life" to a living system when it is most alive, most effective, and most constructively capable in economic, ecological, and human terms. AI involves, in a central way, the art and practice of asking questions that strengthen a system's capacity to apprehend, anticipate, and heighten positive potential. It centrally involves the mobilization of inquiry through the crafting of the "unconditional positive question" often involving hundreds or sometimes thousands of people. In AI the arduous task of intervention gives way to the speed of imagination and innovation; instead of negation, criticism, and spiraling diagnosis, there is discovery, dream, and design. AI seeks, fundamentally, to build a constructive union between a whole people and the massive entirety of what people talk about as past and present capacities: achievements, assets, unexplored potentials, innovations, strengths, elevated thoughts, opportunities, benchmarks, high-point moments, lived values, traditions, strategic competencies, stories, expressions of wisdom, insights into the deeper corporate spirit or soul — and visions of valued and possible futures. Taking all of these together as a gestalt, AI deliberately, in everything it does, seeks to work from accounts of this "positive change core" — and it assumes that every living system has many untapped and rich and inspiring accounts of the positive. Link the energy of this core directly to any change agenda and changes never thought possible are suddenly and democratically mobilized.[43]

Critics of the appreciative inquiry process may believe that the final judgment about "what to do" with recommendations should be in the hands of senior managers. However, the guidance for change that comes from the combined wisdom of diverse people who are close to the problems effectively represents the "wisdom of the crowd." [44] AI maximizes the involvement of individuals to harness this wisdom.

Training in Appreciative Inquiry

The present project was facilitated by the author's attendance at the Foundations Workshop in Appreciative Inquiry in Toronto, Ontario, 18-21 October 2004. The workshop

---

[42] Heuer, *The Psychology of Intelligence Analysis*, Chapter One.

[43] David L. Cooperrider and Diana Whitney, *A Positive Revolution in Change: Appreciative Inquiry,* http://appreciativeinquiry.cwru.edu/intro/whatisai.cfm.

[44] James Surowiecki, *The Wisdom of Crowds* (New York: Anchor Books, 2005), 274.

addressed both theory and practice in AI. Jane Magruder Watkins, co-author of *Apprecia-tive Inquiry: Change at the Speed of Imagination*, was the primary presenter. She has worked closely with David Cooperider, the pioneer of AI, and continues to do so. Ms. Magruder has been an integral member of the National Training Lab (NTL) for many years. "Founded in 1947, NTL Institute, headquartered in Alexandria, Virginia, with a facility in Bethel, Maine, is a not-for-profit educational company of members and staff whose purpose is to advance the field of applied behavioral sciences and to develop change agents for effective leadership for organizations of all variety."[45]

The instructors engaged attendees in the AI process during the course of the four-day workshop. On day one, each participant interviewed another person and then participants broke into two groups to share the interviews and begin to uncover themes. By day three participants were able to formulate a vision of what our learning community would be from the workshop, along with an action plan to stay in contact. Participants collectively produced enactments of "provocative propositions" to arrive at a collective understanding of what was valued. The learning experience was also entertaining. A video showed how the U.S. Navy has implemented AI to develop leadership skills[46]; another illustrated the AI process as used by Canadian National Defence civilian workers.

The goal of group exercises was for each participant to experience the process so that each could become an effective facilitator of AI.

At the workshop, the first phase of Appreciative Inquiry, the "Discovery" phase, was devoted to developing the interview questions. The scope of the present research project was too large for the process to be carried out in all its stages; however, AI proved to be an effective tool for gathering new information and perspectives to continue the relational process.

AI Is Rooted in New Science Paradigms

AI views organizations as living, complex systems and as such, is rooted in the study of complexity and chaos in physics and biology, theories built on the inter-connectedness of the universe. The prevailing scientific paradigm has kept such inter-connectedness beyond our ready comprehension. The following chart captures some of the language or thought processes that characterize old and new visions.

---

[45] See *http://www.ntl.org/about.html*.

[46] See an explanation of this project at *www.nps.navy.mil/cpc*.

Current and Emerging Paradigms	
Current Scientific Paradigm	**Emerging Paradigm**
Newtonian mechanics; reductionism and dichotomous thinking.	Quantum physics and new sciences: self-organizing systems; chaos theory; complexity theory.
We search for a model or method of objectively perceiving the world.	We accept the complexity and subjectivity of the world.
We engage in complex planning for a world we expect to be predictable.	Planning is understood to be a constant process of re-evaluation.
We understand language as the descriptor of reality: "I'll believe it when I see it."	We understand language as the creator of reality: "I'll see it when I believe it."
We see information as power.	We see information as a primal creative force.
We believe in reductionism, i.e., things can be best understood when they are broken into parts.	We seek to understand wholeness and the inter-connectedness of all things.
We engage in dichotomous thinking.	We search for harmony and the common threads of our dialogue.
We believe that there is only one truth for which we must search.	We understand truth to be dependent on the context and current reality.
We believe that influence occurs as a direct result of force exerted from one person to another, i.e., cause and effect.	We understand that influence occurs as a natural part of human interaction.
We live in a linear and hierarchical world.	We live in a circular world of relationships and cooperation.

Source: Watkins and Mohr, *Appreciative Inquiry: Change at the Speed of Imagination,* 8.

Why AI?

AI helps identify the positive core of an organization, whether it be a business, a country, or a discipline, such as intelligence analysis. Inquiry into the positive core is "where the whole organization has an opportunity to value its history, and embrace novelty in transitioning into positive possibilities."[47]

The positive core is identified by inquiry and may be expressed in numerous ways. Some of these expressions were uncovered during the interviews for this project.

---

[47] David L. Cooperider, Diana Whitney, and Jacqueline M. Stravros, *Appreciative Inquiry Handbook: The First in a Series of AI Workbooks for Leaders of Change* (Bedford heights, OH: Lakeshore Publishers, 2003), 30.

Expressions Connoting the Positive Core	
Best business practices	Positive macro strengths
Core and distinctive competencies	Product strengths
Elevated thoughts	Relational resources
Embedded knowledge	Social capital
Innovations	Technical assets
Organizational achievement	Values
Organizational wisdom	Visions of possibility
Positive emotions	Vital traditions
	Strengths of partners

Source: Author's interviews and *Appreciative Inquiry Handbook,* 31.

Identifying such expressions yields energy for change. This is one significant aspect of AI that sets it apart from traditional inquiry approaches.

Principles of Appreciative Inquiry

Whitney and Trosten-Bloom outline eight principles of appreciative inquiry.[48] These principles were applied and their outcomes manifested in the present research, as indicated below.

1. **The Constructionist Principle** — this principle states that reality is socially created through language and conversations, that the words we use create our worlds. A language shift such as changing the "war on terrorism" to a "struggle against violent extremism" and then to the "Long War" illustrates how leaders can consciously try to influence perceptions by their choice of words.[49] We do it every day. By collecting and using the interview subject's words in this project to explain and describe phenomena, the scope of this work is enhanced and transcends the limits of the researcher's perspective. The interview subjects' responses determined the format of this manuscript.

2. **The Simultaneity Principle** — this principle asserts that the act of questioning by itself influences change, or even causes change and is thus an intervention. The present author can attest to this — in the course of the interviews she learned different approaches that have influenced her own work as an analyst. For example, she added

---

[48] Diana Whitney and Amanda Trosten-Bloom, *The Power of Appreciative Inquiry: A Practical Guide to Positive Change* (San Francisco, CA: Berret-Koehler Publishers, Inc, 2003), 54-55.

[49] George Packer, "Comment: Name Calling," *The New Yorker*, 8 and 15 August 2005, 33; Bradley Graham and Josh White, "Abizaid Credited with Popularizing the Term 'Long War',"

more information about known criminals to her analytical products in a way she had not done earlier.

3. **The Poetic Principle** — this principle asserts that organizations are an endless source of study and what we choose to study both describes and creates the world as we know it. The present author hopes that by choosing to study intelligence analysis by focusing on law enforcement practitioners' values, she will influence the way others conduct future studies. By finding "what works" using positive inquiry, the objective of this project is to create change in unanticipated and unexpected ways. Perhaps the introduction of a new concept may influence one reader to work better or smarter.

4. **The Anticipatory Principle** — human systems move in the direction of their visions for the future; the more positive and hopeful the image of the future, the more positive the present-day action. This research project deliberately focused on what interview subjects hoped would continue in the field of crime and intelligence analysis as well as what their deepest wishes were for the future. Theoretically, based on AI research, just by getting the participants to think of these things, the research may help participants to move in the direction of their positive images.

5. **The Positive Principle** — momentum for large-scale change requires large amounts of positive affect and social bonding; positive questions lead to positive change and amplify the positive core. This principle applies to group AI activities, which were beyond the scope of this project; however, the "positive core" identified in this text may help energize the field to expand.

6. **The Wholeness Principle** — wholeness brings out the best in people and organizations; bringing all stakeholders together stimulates creativity and builds capacity. Although this project was limited to questioning individuals one-on-one, the production of this manuscript for the larger audience of analysts, managers, and policymakers from various, often unconnected realms, is in itself an attempt to create wholeness out of what has been traditionally viewed as parts.

7. **The Enactment Principle** — acting "as if" is self-fulfilling; to make the change we want, we must be the change we want. By creating this project conceptually and carrying it out, the author enacted the changes she wanted to see. She believed it was possible; she did it. Enactment can be implemented on a wider scale by any reader interested in influencing change.

8. **The Free Choice Principle** — free choice liberates power; it stimulates organizational excellence and positive change. It appears that analysts who have freedom enjoy it and say it contributes to their growth as professionals. With respect to this specific research project, it was the author who chose what she would do. The freedom she had to design and carry out this project provided the energy and excitement necessary to carry it out from beginning to end. The participants all volunteered to be interviewed and thus this was all a free-will effort. AI processes are not coercive.

The Interview Process

Fifty-two people were formally interviewed in order to learn more about the context of success in law enforcement analysis. A large number of analysts/experts agreed to be interviewed but many were so busy that interviews could not be coordinated in the time frame allotted for this project. A number of the participants, but not all, were acquainted professionally with the researcher prior to the research. (Thirteen out of fifty-two had never met the researcher.) The participants were chosen based on their contribution to the field or on the dimension of expertise and/or diversity they offered to the study. A few individuals asked to participate and were included if their position was appropriate for the study. Since no definitive count of the number of law enforcement analysts exists, there is no way to compare the size of the sample of persons interviewed to the total population of analysts/experts.

Subjects signed a consent form prior to each interview. Although many analysts have agreed to be identified in this project, some have chosen to remain anonymous. Therefore, throughout this manuscript, some quotes are attributed to an anonymous source.

Most interviews were conducted by telephone and typically lasted about forty-five minutes. Nine interviews took place in person.

The Participants

Participants by Category	
16 Crime Analysts	11 Intelligence Analysts
8 Both Crime and Intelligence Analysts	8 Crime Analysis Academic Researchers/ Developers
3 Intelligence Analysis Academic Researchers/Developers	2 Command Staff with Analysis Experience
4 Others Relevant to the Study	52 Total Number of Subjects

At least fifteen of those interviewed had analytical experience or expertise in analysis at the federal or national level of law enforcement, and seven of those interviewed had analytical experience or expertise in analysis at the state or regional levels. Twenty-three had analytical experience focused on analysis at the municipal level of law enforcement. Four were Canadian analysts, and two had previously worked in analytical positions in Australia. Three individuals had military intelligence analysis experience. Nine of the participants were analysts who held a supervisory position specific to criminal analysis or intelligence analysis at the time of the interview. At least thirteen of those interviewed were actively involved in training analysts. Ten were teaching college-level courses in analysis. Eight of those interviewed had written books related to law enforcement analysis; they and eight others interviewed had contributed to the literature in the field. Their formal education was in very diverse fields, as shown below.

Educational Background of Interview Subjects	
Accounting	Human Resources
Anthropology	Journalism
Biology	Law Enforcement
Business Administration	Library Science
Chemistry	Management
Computer Science	Marketing
Criminal Justice	Mass Communications
Criminology	Mathematics
Economics	Political Science
Education	Psychology
Engineering	Public Administration
English	Public Policy
Geographic Information Systems	Sociology
Geography	Theology
Government Administration	Urban Planning
History	Vocational Education

The Interview Schedule

The participants were asked the following set of questions.[50]

- To start, I'd like to learn how you came to work in the field of law enforcement intelligence analysis. When did you start working in the field and what attracted you to this field? What keeps you working in the field, or, if you have left it, what kept you working in it as long as you did? What about law enforcement intelligence analysis makes a **difference** to you?
- Tell me a story about a time that stands out to you as a high point in your work as an analyst — a time when you felt energized, passionate about your work and most effective — a time when you were able to accomplish more than you imagined.
- What do you value most — specifically about yourself and LE intelligence analysis as a field?
  - ❏ Without being humble, what do you value most about **yourself** — as a human being? What are the most important qualities or strengths you bring to this field?
  - ❏ What is it about the nature of the work **you** do that you value the most? What is most interesting or meaningful?
- To grow as a field, law enforcement intelligence analysis must learn to "preserve the core" and be able to let go of things that are no longer needed. In transforming this field, what are **three things** — core strengths, values, qualities, ways of working — you want to see preserved and leveraged as we move into the future?
- Fast forward ... it is now 2010. What would you envision for the field of law enforcement intelligence analysis? If you had a magic wand, what would you do to change the field of law enforcement intelligence analysis?

---

[50] Note that law enforcement intelligence analysis also refers to crime analysis in this interview.

Rationale for the Questions

The first question, asking the participants how they came to this work and what it is that keeps them working in it, helped set the stage for the subsequent questions by focusing on *attraction*, a positive emotion. A number of individuals came to this work by chance: they saw a posting, applied, and got the job. Some were already in law enforcement and desired a change. Almost all have stayed in the work because they felt they could contribute to society in one way or another — they felt there was meaning in this work on a personal level. No one indicated that they did this work for the money. Since this first question was open-ended, the information elicited from each participant varied more than in the other questions.

The second question, asking participants to *tell a story* about a high point in their work, one in which they felt they could accomplish more than they had imagined, reflects a core process in AI. The story, in AI theory, provides energy and uncovers those themes that contain common threads upon which we can build relationships and become energized. In this project, the AI process was diluted because the author was the story's only listener. As AI is usually carried out, a group collectively locates the themes in the stories. It is in the group sharing of stories that the AI process moves forward. Nevertheless, some of the subjects' stories became the success stories of this book. *The high point stories were most often the things that work.* In that regard, using AI as an interview format was very effective.

The next questions, asking analysts what they value about themselves and what they value in the actual work they do, helped direct the participants' attention to their own participation in the work process. What positives they bring to their work and what actual aspects are most enjoyable were concepts that a significant number of those interviewed had never thought of before. The qualities they valued in themselves are the things that work on an individual level. The aspects of the work they like most help clarify what the work actually is — our understanding of this detail is still evolving as the field grows. The answers helped the author articulate these concepts with a greater clarity later in this text.

The next question, what are *three things* — core strengths, values, qualities, ways of working — you want to see preserved and leveraged as we move into the future, is the most important question in that it uncovers "what works" in law enforcement intelligence analysis. What was most surprising to the author was the number of analysts who could not think of three things that work. One said nothing works but then qualified his answer by indicating that nothing works the way he would like it to work.

The things that work as perceived by the people who do this work are very important to know for several reasons. First, we know what to build upon if collectively we agree on the articulated value. Second, we do not discard or mistakenly modify something that is currently effective. Third, we better understand what it is we do right and try to take those lessons to other areas of need.

The last question, about the future, is designed to help create the vision we would hopefully move toward as we seek to grow a profession, and as a nation committed to developing new measures to ensure public safety. "The heliotropic hypothesis suggests that people and organizations, like plants, will move in the direction that is most life-giving."[51] Just as plants turn and move toward the sun, so do people turn and move to images that inspire them.

For the author, the interviews were a highlight of her professional life. They were energizing, pleasant, very informative, and she learned something in every encounter. When a participant would go off onto a negative tangent, the questions themselves provided a return to considering the positive. A number of the participants seemed to enjoy the interviews as well.

Data collected were shared with participants and others through an online "blog." It was hoped that this would provide an interactive forum for participants, but since they were not as committed nor invested to the process as the author herself, it received minimal attention. It did open doors to the researcher, nevertheless, as she was subsequently invited to join the Police Futurists International after a member of that association visited the blog site.

Limitations of This Project

The scope of this project does not address the many concerns our nation faces as we move toward using intelligence to inform law enforcement strategies. Civil liberties, constitutional rights, and privacy concerns are fundamental aspects of the law enforcement context that are discussed in other resources. They are too complex for this project to address. Such major concerns related to the development of intelligence analysis in law enforcement should not be forgotten nor dismissed in the dialogue and debate on this subject.

Although this book covers many elements of the law enforcement intelligence analysis field, it is selective in that it used only those things that fell under the umbrella of phenomena expressed by the interview participants. The interview data were supplemented with research into what is known about success in the categories defined by the participants. This work is limited by the boundaries provided through the interview process and outcomes. At the same time, however, interview subjects often demonstrated that their innovations exceeded the boundaries established by precedent in law enforcement intelligence analysis. Finally, this document gains strength, in AI terms, by focusing on themes of value to those with experience and expertise.

A possible shortfall of this product is the minimal participation by analysts at the federal/national level of law enforcement. Out of fifty-two persons interviewed, only nine had that level of analytical experience. Several more had worked at the federal level in related capacities but were not working analysts. Only seven had regional or state level experience. Although more analysts as these levels were invited to participate and some replied that they were eager to do so, they did not carry through by scheduling an interview. This outcome tended to move the focus of this project to that of the local level of law enforcement.

---

[51] Watkins and Mohr, 135.

One explanation of this bias toward local efforts has to do with the restrictions placed on sharing at the higher levels of government. For the most part, local-level law enforcement analysts are directed by their job descriptions to network and share information outside of their respective jurisdictions, and many have the authority to do so without restriction, except to use sound judgment regarding what can be shared with whom. This is not the case with many analysts at other levels of law enforcement. Much of their work is case-specific and sharing is restricted by stricter regulatory guidelines. Outcomes of investigations and prosecutions depend on their maintaining secrecy. Thus, they could not as easily obtain organizational permission to participate in a study about analysis.

Another limitation of this project is that some complicated concepts and topics were, by necessity of time and space, reduced to a paragraph or two. In such cases, one or more links to additional information are provided to assist readers in learning more.

Next

The next seven chapters are organized around the general categories of the "things that work" as identified in the interview process. Therefore, the participants formed the structure of this text, limiting the author's task to describing in greater depth the things that already work. Chapter Ten discusses analysts' wishes for the future. By the *Anticipatory Principle* of AI, it is expected that, by capturing the collective vision of over 830 years of law enforcement research/analytical experience, we will move into the direction of a future desired by active and influential workers in the field.

Chapter 3

WHAT WORKS: INDIVIDUALS

My job is to turn information into knowledge.

I think I was a bloodhound in another life. [52]

Systems-thinking business expert Peter Senge writes in *The Fifth Discipline: The Art and Practice of the Learning Organization* that by seeing information in both broad and detailed patterns we can respond to the challenge of complexity and change.[53] In order to develop the analytical capacity of law enforcement, we need to look in detail at how it works at the ground level as well as in broader theoretical, conceptual terms. We begin by looking at the analysts and field experts as individuals.

Why Study the Individual Intelligence Analyst?

To best understand something, one must have real-life experience with it. Too often in law enforcement intelligence analysis, no one thinks to ask the analysts what it is they do, how they do it, what they need, and what they think matters. Most often, their supervisors have never been analysts. Staff is sent away for distant training to become analysts, but managers are not sent away to understand what the analysis requires, what it is supposed to be nor how to do it. The analyst often has to beg for the tools he or she needs, as well as for even the most basic information necessary to do quality analysis.

> Dictionary definitions frequently offer verbal substitutes for an unknown term which only conceals a lack of real understanding. Thus a person might look up a foreign word and be quite satisfied with the meaning "bullfinch" without the slightest ability to identify or describe the bird. Understanding does not come through dealing with words alone, but with the things for which they stand. Dictionary definitions permit us to hide from ourselves and others the extent of our ignorance. [54]

This chapter introduces readers to law enforcement analysts through some of their "high point" stories, and through the qualities and abilities they value as individuals. We start with an overview of the analyst's role in law enforcement, including the skill sets and characteristics needed by individual workers. The chapter concludes with "what works" in the context of the individual analyst.

---

[52] Quotes from interviews.

[53] Peter Senge, *The Fifth Discipline: The Art and Practice of The Learning Organization* (New York: Currency Doubleday, 1990), 135.

[54] *Language in Thought and Action*, 33, quoting H.R. Huse.

The Stories

Because narratives provide context for understanding, the stories of analysts' "high points" may give readers a sense of what analytical work "looks like" when it works. The stories as reproduced here are no substitute for actually hearing the stories and the energy behind them; however, the basic work stories and "archetypal stories" delineate some of the aspects of analytical work that impart energy to participants.

The Basic Types of Work Stories

Many analysts had high points that involved four main categories:

- **The Identification of a Crime Series Story:** A crime analyst is the first one to notice an existing pattern of crimes in which the same perpetrator(s) seems to be responsible — a crime series — and the series does, in fact, exist.
- **The Pieces of Information Turning into a Big Case Story:** An intelligence analyst gets boxes of information, sometimes CDs full of information, and sorts through all of it (analyzes it) to uncover information that leads to an even bigger investigation.
- **The Prediction Leading to Arrest Story:** A crime analyst makes a prediction regarding the next likely time and place a serial criminal will offend and the offender is apprehended based on the analyst's accurate prediction.
- **The Successful Investigation Leading to Prosecution Story:** An intelligence analyst supports an investigation through appropriate analysis and visualization of data, creating relevant reports and graphics; the analyst's work is used in court to help successfully prosecute the targets of the investigation.

The "Archetypal" Stories

Within the analysts' "high point" stories certain common themes emerged that transcended the details of their work. In the present author's view, the story types are: collaboration, support, discovery, creation, invention, influence, achievement, recognition and impact. These themes are intertwined multiple times in a majority of the individual stories.

The Collaboration Story

Aspects of collaboration came up in many high point stories. Collaboration can mean working in teams with clear but ordinary procedures and routines in place that repeatedly result in extraordinary results. It can mean working on diverse teams where brainstorming ideas with a variety of perspectives energizes participants. Linking cases with physical evidence across jurisdictions is another form of collaboration. Task forces are a form of collaboration often found to be effective. Managers who carefully select the analysts on their teams tend to be enjoy collaborating with them.

For analysts, collaboration with enforcement officers can take several forms. Associating with the 20% of the officers that do the work, "the elite," is a form of collaboration that helps an analyst. Attending officer training with officers is another activity that helps build relationships for future collaboration. Inclusion in briefings and meetings with officers and investigators, going out into the field with them, and conversely being able to train them to gather the data needed for crime analysis is a form of collaboration that works. In sum, building upon analysis in iterative fashion by getting feedback from officers and investigators, and thereby changing one's analytical perspective, is an approach that works.

Here are some more specific examples:

- Multi-agency collaboration is exciting to a number of analysts. Teamwork between many analysts and agencies, such as the local law enforcement agency, the IRS (Internal Revenue Service), the Federal Bureau of Investigation (FBI) (for behavior analysis), and RISS (Regional Information Sharing Systems) analysts to complete a detailed picture and to prove a suspect's motive for murder in Kansas City, MO, is an example of such a multi-agency collaboration provided by Sergeant Gregory Volker.

- Cooperation between law enforcement agencies outside their own jurisdictions in tackling problems can help put pieces of a puzzle together. An example comes from a remote location in Canada where empty boxes of stolen goods had been found for years by railroad personnel; François Dubec, railroad intelligence analyst, along with a number of law enforcement agencies from Canada, collaborated and ultimately identified a Hell's Angels fencing ring.

- The cooperative and collaborative effort of working on the National Criminal Intelligence Sharing Plan was a high point for Lisa Palmieri, president of IALEIA.

- In the mid-1990s Greg Saville collaborated with the Sydney Olympics Foundation over a number of weeks, planning with designers and architects to help them think about crime risks and successfully apply Crime Prevention Through Environmental Design (CPTED) theories in the real world.

The Supported Story

Another theme in the high point stories involved being supported by others. This is a less common theme, yet it is important. Noah Fritz says he was able to build a true infrastructure for crime analysis in Tempe, Arizona because the police department had supportive chiefs. Having support means you have access to information, that you become integrated into the agency, and that you get the training and equipment necessary to be successful. The Tempe PD subsequently has produced a number of analysts who went on to high-level jobs influencing the growth of crime analysis.

Others said they were supported when trying to implement new ideas, such as geographic profiling or environmental scanning. Shelagh Dorn indicated that her management team consists of progressive thinkers who are extremely supportive of analysts in the relatively new Upstate New York Regional Intelligence Center.

The Discovery Story

Discovery involves noticing something that one has not noticed before. It is especially prized when one is the first to notice something that is valuable to know. Discovery in analytical work may consist of getting information that initiates new investigations, noticing how situations impact on crime rather than people, and learning how to focus on the "important stuff" in the midst of an information glut. Discovery is the core of much of analytical work and was part of a number of high point stories that mainly focused on "impact."

Here are some examples:

■ Toronto sworn officer/analyst Tyrone Skanes discovered the identity of two people who got away at a murder scene based on data analysis. Peel Regional Police had already arrested two persons and needed to know the identity of the two outstanding suspects. He was able to give them two names plus two lists of people to talk to — those who would and those who would not cooperate. It took Peel officers three months to develop information leading to arrest, but he was correct. He knew who to look for, where to go in the data banks, and in using field contact cards (called 208s in Toronto), he assessed a wealth of information.

■ NW3C Training Director Mark Gage experienced high points in discovering new areas for training in computer crime, white collar crime and terrorism, and economic crime in households.

■ Mesa, Arizona analyst Peter Garza discovered a link between an auto theft case and a burglary case; subsequently the burglary suspect admitted to stealing the automobile. Discovering a link between two different crimes does not happen unless one looks for it.

The Creation Story

Creativity in analysis often involves the synthesis of a lot of information to create a useful analytical product. Analytical work is creative — the outcome of analysis in the form of analytical reports requires that an analyst make many decisions along the way as to what should go in the product and what should not. Creativity extends to developing programs that help "grow" the field of analysis.

Some examples:

■ After two weeks on the job working for the U.S. Attorney, Western District of New York, intelligence analyst Ronald Wheeler participated in the litigation of the Lackawanna Six; he created presentations for court.

■ Karin Schmerler designed a program at the COPS office called "Problem Solving Partnerships" which offered "funding only for analysis" to four hundred projects — some people won POP awards. She liked helping "grow" projects.

■ Lois Higgins developed the Florida Department of Law Enforcement (FDLE) Analyst Academy and hand-picked the first class to tweak the curriculum, which was based on the Community Model from the Counterdrug Secretariat. She adapted it to fit Florida, in order to get counterterrorism funding from the federal government; she adapted the model because it was "blessed by the Feds."

■ Paul Wormeli was the originator of the Integrated Career Apprehension Program which identified career criminals and put more attention on prosecution. This program in turn funded agencies to create crime analysts.

The following story, from Crime Analyst Peggy Call, Salt Lake City Police Department, illustrates how creative an analyst can be in her thinking, and how persistence pays off in following the information trail to wherever it leads.

In the spring of 2003, we had a recurring (myth or truth) series of what the public and some in animal organizations deemed to be "cat mutilations." Some people believed that these cat mutilations were "human caused" as the cats were "cut with precision." After much research, involving drought/rainfall variables, media stories, a 10-year search for similar incidents — which asked our partners with animal and humane societies to go back through their files to try and provide further details of these incidents — and with the general public claiming that this was all due to a serial killer in our midst, we took the novel approach of looking into every type of scenario possible that could lead us to the cause of these "cat mutilations." Police don't normally get involved in animal matters, unless, of course, there is animal cruelty involved, but our department felt it was in the best interest of the public to address their fears and perceptions on this issue.

After duly searching all reports of disturbances, of animal cruelty, and identifying sex offenders in the target area, we determined that there was nothing that was pointing us in the direction of any "escalating behavior"; nor was there anything that indicated cats were being trapped and abused/killed; nor were there any neighborhood complaints of "screaming cats." We started adding some other scenarios to the investigation; we kept everything on the table until proven or discounted.

I used GIS (Geographic Information Systems) to pinpoint the area where these occurred. Year after year, they occurred in the same area. At the same time, Aurora, CO was experiencing similar incidents and we found, through GIS, that Denver's drought pattern was the same as ours. I also did some research on coyotes, foxes and Great Horned Owls as perpetrators (much to everyone's dismay and claims that "there were **no** Great Horned Owls in Utah"). I was able to illustrate and examine possibly relevant aspects of migration, breeding, drought, pattern of incidents, rainfall, and more for the entire area of Aurora and Salt Lake City.

My question every day was "Why in both of these locations?" What is similar about them? Is it possible to have a serial killer operating in both cities? As we looked at the time of year, rainfall, and physical geographical features of those specific areas, the similarities were so strikingly similar, we knew we had to keep looking at the "Why."

Armed with this information, we contacted the Division of Wildlife Resources. They were extremely helpful in showing us about animal life, scat patterns, and the like, and actually took us up on horseback to the key area and pointed out where coyotes had marked their area. They also located a fox den close to the key area and removed the foxes. They then, through labs, determined that the foxes there had cat hair/skin in their claws and at that point, we had one cat carcass (portion) left in a neighborhood and it was determined, through labs, that the cat had fox hair in its claws!

The fox den was removed. I've not had another incident since.

Bottom line? Think out of the box. Keep digging. Be persistent. You don't know until you know.

Examples of Analytic Graphics for the Mutilated Cat Case

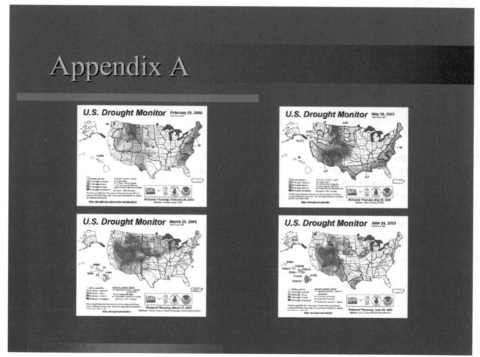

Between February of 2002 and July of 2003 Utah and Colorado experienced a move from abnormally dry weather conditions to extreme drought even during winter months. The pattern is also observed to move from east to west from Colorado to Utah, which may account for the large number of cat deaths in Colorado. This may be a good barometer for things to come for Utah, associated with animal predation and the competition for resources.

Source: U.S. Geological Survey.

Appendix B

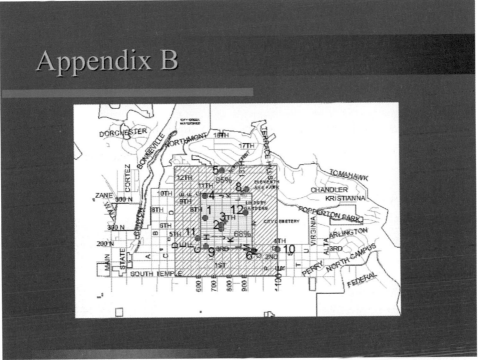

Portion of Salt Lake City, showing location of mutilated cat carcasses. The geography of the Salt Lake City area is similar to that of Aurora. In Colorado, there is a ravine/creek dividing two areas of land, much like Capitol Hills and the Avenues in Salt Lake. The area encroaches on wildlife areas. As well, the Colorado area is experiencing extreme drought. The Avenues has a large area of wilderness to the north and east as well as open parks and large cemeteries. This type of development provides easy cover for wild animals that have become accustomed to a human presence and are accustomed to easy food sources when other traditional means of feeding are threatened.

Source: Peggy Call.

The Invention Story

Analysts are often inventive in figuring how to do their work despite numerous obstacles to success. Several of those interviewed are true "inventors." Here are two of them.

■ John Eck was selected to head the problem-oriented policing (POP) project in Newport News, Virginia, which ran from 1984 to 1987. He thought POP was "crap" at first. Bill Spellman was hired as an assistant. Because this was a grant-funded project, they went to Newport News weekly. Chief Darryl Stevens put together a team of every rank/unit, about 12 people in all, to work with them. The project could be summed up at first with the question, "What is POP and how are we going to implement it?" To address this question, Bill and John created the "SARA" (scanning, analysis, response, assessment) process that is central to POP (see Chapter Five); it was crafted and conceptualized for this project. Three important case studies grew from this work: a downtown prostitution/john problem, a theft from unsecured parking lots problem, and a suburban chronic burglary rate problem. *U.S. News and World Report* did a three-page spread on their work in Newport News as an example of good Police Department work.

■ Vancouver, Canada detective and Ph.D. student (at the time) Kim Rossmo had an idea for an algorithm while riding on a train in Japan in 2001. The project required developing a computer program — ultimately resulting in what is now called "geographic profiling." Nine months after the idea in the train, he got the program working. At 2AM one morning he ran the program on the very first crime series — and it worked. It gave him a huge feeling of satisfaction when the time investment and the gamble paid off.

The Influence Story

Influencing others was another theme that appeared in analyst high point stories.

Here are some examples.

■ Bryan Hill influenced decisionmakers in the Phoenix PD to use ArcView GIS software and to hire civilian analysts (even though he was a uniformed (sworn) officer).

■ Dr. Ron Clarke's 1976 research work "Crime as Opportunity"[55] was published and went against conventional views — no one else had argued the point that crime is situational. He likes to persuade people to his point of view and works on the fringes of criminology studying the situational determinants of crime.

■ One analyst influenced change when he asserted himself and broke away from the perceived ways of doing things, after agents started coming to him directly rather than going to the de facto gatekeeper — the more senior person who was not his supervisor but who was perceived as such. This analyst makes "house calls" and believes that to support agents we should go to them rather than vice versa.

■ Dr. Jerry Ratcliffe values reaching someone who works in the field through his presentations, by inspiring them and influencing them to improve.

■ Dr. Rachel Boba values when people "get it" during training, when she shows examples of what crime analysis does and people can see new ways of looking at crime. She asks, "how to bring analysis to entire organizations" — and sees the answer as a participatory evolutionary process.

■ In 1989, for several weeks, RCMP analyst Angus Smith interviewed a downed pilot who had landed in Canada and who worked for the Cali (Colombia) Cartel. He subsequently wrote a report about changes in the cocaine trade based on the debriefing. His strategic report seemed to "light a fire" under people and he was taken seriously. Holes in radar were detected and North American law enforcement organizations responded to the report by altering the way they were addressing the Cali Cartel drug trade.

---

[55] Ron Clarke and others, "Crime as Opportunity," Home Office Research Studies No. 34 (London: Her Majesty's Stationery Office — HMSO, 1976).

The Achievement Story

For some analysts, the high point is achievement. Achievement comes in different guises: being a leader in the field, earning the respect of others, being viewed as an authority via one's achievements, and being hired in a competitive job because of one's achievements. Writing and getting published, being offered teaching jobs, the ability to do all sorts of work well — these also are part of achievement.

■ Crime Analyst Metre Lewis says that now that she has become a Florida Department of Law Enforcement-certified Crime Analyst, she feels that, with 300 hours of training, she is now a professional with the authority behind her to make recommendations, instead of giving her end-users a minimum (what they ask for). Her ability to do in-depth assessments is an achievement that is quite rewarding.

■ One of Retired NYPD Captain Joseph Concannon's high points was a NarcStat meeting wherein he was quizzed on the successes in his precinct. It was noted that his regular officers executed more search warrants for narcotics than the narcotics unit. This was because the approach they took to minor/small crime arrestees, who were treated like human beings — given snacks or meals — as they were debriefed by precinct officers, worked better than the narcotics unit's approaches.

■ A high point for (former Australian) intelligence analyst Howard Clarke occurred while he was working with the British Columbia provincial police in Vancouver, in 1996-1997. During a one-year-long investigation of Asian organized crime with ties to Los Angeles and Seattle, as well as FBI involvement, he valued the experience of taking an ambiguous assignment of analyzing two boxes of material and turning the result into a fruitful investigation through the interpretation and identification of new leads and targets.

The Recognition Story

Recognition takes the form of awards, interagency recognition of the value of one's work, peer recognition, and media attention. Being recognized by peers is most important since public recognition is seldom possible in a closed community like that of law enforcement intelligence professionals.

Here are some examples.

■ Robert Heibel's high point was recognition for his Research and Intelligence Analyst Program at Mercyhurst College. In 1996 NSA sent out a team to evaluate the program for its eligibility for an equipment grant, appraised the program and decided to support it — the program had "crossed the threshold" and gained acceptance by the national Intelligence Community.

■ One of David Jimenez's high points was his IALEIA award in 1998 for Border Patrol/ INS work, a recognition by the profession for his analytical work on a major smuggling operation investigation.

- One of Steve Gottlieb's high points was national coverage on NPR and NBC evening news of a bank-robbery series arrest made by officers who had used his forecasting methods after attending his analysis course. The resulting acclaim for crime analysis and increased public awareness was very rewarding to him.
- Mary Garrand created an intranet Crime Analysis (CA) page, which officers could access in their patrol cars. An officer on the midnight shift "out of the blue" sent her a message about how great the CA Unit was because the information on the page was very helpful. (See box below.[56])

To the Alexandria, Virginia Crime Analysis Unit:

This is a notice of commendation and appreciation for Officers P. Alvarez, R. Enslen, G. Hillard, and the Crime Analysis Unit.

In the beginning of March, I put together an investigative package on a wanted sex offender residing in my group's assigned sector. The information on the subject was discovered from the newly developed Crime Analysis Unit's Sex Offender Information page located on the Mobile Data Terminals. This package was copied and distributed to officers in my group for follow-up investigation that would lead to an arrest. One of these copies was forwarded to Sgt. W. Mayfield and a request was made for the mentioned officers to check on the case. These officers had been working a repeat offender's detail, initiated by Lieutenant D. Gaunt. The suspect in this case was wanted for failing to register as a sex offender, a felony warrant issued on 12/4/2004.

After several attempts and a surveillance operation, on April 1 these officers were able to locate and arrest the fugitive at one of the addresses included in the investigation package.

I wanted to thank these officers for taking the extra time and effort in locating this felon. I also wanted to pass along a note regarding the effectiveness of the Crime Analysis Unit's MDT page and to let them know that their information gathering and distributions do get noticed and often lead to taking criminals off the street.

The Impact Story

Making a difference in the real world, having some impact, is a consistently mentioned high point. Preventing a potential victim from victimization is a primary motivator for several analysts. Predicting where and when an offender might crop up, tying various events into a single pattern, helping officers find offenders, and connecting crimes and criminals all have impact for analysts.

---

[56] Courtesy of Mary Garrand, Crime Analyst Supervisor.

Here are some examples:

- For Shelagh Dorn, analyst and analyst supervisor in the Upstate New York Regional Intelligence Center, impact includes opportunity to set up an analytical unit, being able to build it with advanced technology and producing results rapidly.

- For Sergeant Mark Stallo of the Dallas, TX PD, developing the concept of the crime analyst for civil service, including how to advertise and interview for such positions, had impact. He was able to use what he had seen to be effective in his experience, his knowledge of software requirements, and his awareness of what qualities had led to success for analysts he had supervised, to help determine the qualifications of future analysts in his agency.

- For Chief Tom Casady of the Lincoln, NE PD, the November 2002 arrest of Richard Eugene Merritt for indecent exposure was a high point centered around impact. Taking a pedophile off the street with the help of a young boy who had suspicions, and a license plate number, and taking the boy seriously, was part of the process that led to further investigation and arrest.

- Joseph Regali, who works in the New England RISS, found out that six to eight of the persons targeted for a case he worked on for a year (100-150 hours) were now in jail, that his work made a difference, that his product format was used for the purpose it was intended — prosecution — and that he was able to provide more than investigators could do on their own.

- Mary Garrand worked on a series of robberies of banks at strip malls and made a successful prediction. She used CPTED (Crime Prevention Through Environmental Design) techniques and crime prevention techniques in forays out into the field to assess environments herself. The robber was caught with the robbery note in his pocket where and when predicted.

- Lorie Velarde also likes to do environmental surveys to find the unconscious environmental clues that criminals react to when choosing places/victims, and feels rewarded when good arrests result from her work.

- Chris Delaney enjoys the ability to link what he is doing to results, and seeing the significant drop in homicide rates in Rochester, NY was a high point in the process of working on Operation Cease Fire.

Several analysts found teaching others very rewarding. One said that providing crime analysis training on an in-service basis to officers was a high point; the training focused on improving data collection or analysis and data integrity.

What Analysts Value in Themselves

Characteristics

accessible	finds meaning in the work	persuasive
articulate	future-oriented	practical
committed	good listener	responsible
convincing	good memory	risk-taker
cooperative	helpful	self-aware
creative	humility	self-confidence
data-hungry	independent	self-motivation
dedicated	intelligent	sense of justice
desire to contribute	intrinsically motivated	sincere
desire to learn	intuitive	strategic thinker
detail-oriented	knows territory	strong work ethic
disciplined	learns from mistakes	tenacious
enjoys diverse viewpoints	love of people	truthful
enjoys novelty	loyal	unafraid of criticism
enjoys variety	open to ideas	unorthodox thinker
enjoys work	passionate	willingness to work extra time
fair	patient	work experience
faith in God	perpetual student	

Abilities

to direct	to engineer	to provide structure to concepts
to distinguish concepts	to finish worthy projects	to see potential
to adapt	to follow through	to share
to adapt products to audiences	to follow trail	to simplify the complex
to ask questions	to help group reach consensus	to solve problems
to assert	to identify needs	to synthesize
to break things down	to juggle conflicting concepts	to understand limitations of technology
to build	to keep to a timetable	to visualize
to challenge	to know when enough is enough	to work in teams
to deliver to consumer	to make connections	to work with others
to dig deeper	to prioritize	

DEFINING THE INDIVIDUAL ANALYST

Affiliation with Associations of Law Enforcement Analysts

Law enforcement analysts have formed two associations, IALEIA and IACA. These associations have become the de facto authorities that define job performance standards for law enforcement analysts, as we have neither national standards nor alternative mechanisms of oversight. Individual analysts in both the IALEIA and the IACA have developed standards for certification processes for their respective associations, and to apply those standards, a peer group of analysts defines desirable individual characteristics for the population of analysts.[57]

Certification is always at the individual level. Individuals demonstrate their abilities and achievements through a defined process. Thus, a standard is set for the individual analyst who chooses to go through the process. As of 2005, over 200 individuals worldwide have met IALEIA's Society of Certified Criminal Analysts (C.C.A.) standards and are certified as criminal analysts.

The IACA expects to administer its first certification in 2005 and the title of those certified will be Certified Law Enforcement Analysts (CLEA). [58]

The IACA Certification Program is a result of the desire to reach the following six goals:

1. To recognize the professional abilities and accomplishments of individual law enforcement analysts.

2. To promote and encourage professional development by individuals in the field of law enforcement analysis.

3. To provide the employers of law enforcement analysts a reliable measure of professional competence.

4. To provide employers of law enforcement analysts with a basis on which to establish position descriptions.

5. To promote the profession of law enforcement analysis to police chiefs, administrators and the entire criminal justice community.

6. To better define law enforcement analysis as a legitimate and unique career. [59]

---

[57] Courtesy of Mary Garrand, Crime Analyst Supervisor.

[58] See *http://www.iaca.net/Certification.asp*.

[59] See *http://www.iaca.net/Certification.asp*.

The Individual Analyst in the IC

We can infer that the analyst in the national Intelligence Community faces some of the same challenges as an analyst in law enforcement and requires the same attributes, if we examine the perspectives of the leading IC experts. Mr. Adrian (Zeke) Wolfberg of the Defense Intelligence Agency describes the overriding importance of *how an analyst thinks* over and above *what an analyst knows* in the passage below.

> When we think about the all-source analyst, it is usually in terms of things they do and the knowledge they possess. Analysts identify a problem, find information about it, think about the information, reach a conclusion and then write a report addressing the problem for a policy-maker or decision-maker. And there are training and educational programs that target these tasks and improve their knowledge of facts. However, if we think about the analyst in terms of how they think — which is the fundamental "crown jewel" of analytical work — then we immediately must acknowledge its importance over what they know. [60]

> These analytical attributes include: innovation, synthesis, learning, questioning, pattern recognition, adaptation to uncertainty, visual thinking, experimentation, metaphors, nonlinear systemic thinking, focus on what is unknown or unknowable, asking "What If?," and building and working with multiple frameworks. [61]

Next, Mark Lowenthal, then of the CIA, describes desirable analytical qualities.

> We do not train analysts well. Their early training is good and then they are forgotten. No one looks out for careers (career management). Many analysts become disaffected and leave. There is a continuing need to know everything about everything on a moment's notice and there are no analytical reserves regarding human resources.

> There are not enough mid-level analysts and we have to deal with collaboration vs. competitive analysis. Why do people become analysts? They become analysts because of an intellectual curiosity, they like to share knowledge with other people, they tend to like to write even if it's painful, and they are attracted to public service. [62]

The findings of this study echo these passages. Law Enforcement (LE) Analysts are intellectually curious, desirous of sharing what they know, and of making a difference for the public by their service. They are curious innovators who explore many questions down paths that take them in unexpected directions. They adapt, experiment, and learn

---

[60] Adrian (Zeke) Wolfberg, "To Transform Into a More Capable Intelligence Community: A Paradigm Shift in the Analyst Selection Strategy," National Defense University, National War College paper, 4. URL *http://www.ndu.edu/library/n4/n03AWolfbergParadigm.pdf.*

[61] Wolfberg, 20-21.

[62] Mark Lowenthal, "The Future of Analysis," Keynote Speech, Findings: Fifth Annual International Colloquium On Intelligence, Mercyhurst College, Erie, PA, 9-11 June 2003, 16.

continually. Writing and visualization of information are central to their work. They synthesize information from a variety of sources, limited only by their imaginations, resources and agency hierarchies, to develop deeper understandings of problems. What the analyst knows is less important than how he or she thinks — no one in this study emphasized knowledge alone as valuable. Experience, yes. Those interviewed affirmed that they valued their ability to think.

Still other elements of LE analysis mirror what we know about analysts in the national Intelligence Community. Pattern recognition is another core aspect of LE analysts' work, for example, especially in local law enforcement wherein many analysts are sifting through large data sets to uncover patterns on a daily basis. LE analysts also experience the same frustrations as analysts in the national Intelligence Community. Managers do not nurture them and their career paths are very limited. Training is often only provided at the beginning of their careers. They often have too much to do and no backups to cover them or supplement them — some operate alone and all of the analytical work of their agency is up to them.

One difference in comparing LE analysts to those in the national Intelligence Community is that no LE analysts spoke of competition as a problem. Collaboration is the rule in LE analysis. Perhaps because analysts are scattered (very rarely are there more than a few analysts working together), competition is less likely and collaboration is necessary to build any sort of LE analytical community.

WHAT WORKS AT THE INDIVIDUAL LEVEL

Interview subjects identified individual aspects of analytical work that contribute to success. This chapter concludes with "what works" on an individual level as reported by those interviewed.

What Works: Desire to "Catch the Bad Guys"

The ability to help "catch the bad guys" keeps analysts in this work. One analyst said, quoting Ernest Hemingway, "Certainly there is no hunting like the hunting of man and those who have hunted armed men long enough and liked it, never really care for anything else thereafter." [63]

---

[63] The quote appeared in the April 1936 issue of *Esquire*. It was the first line of an article titled "On The Blue Water: A Gulf Stream Letter." URL *http://www.lostgeneration.com/hem-faq.htm#hunting.*

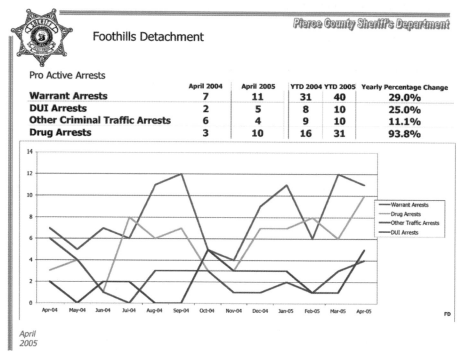

Pro Active Arrests

	April 2004	April 2005	YTD 2004	YTD 2005	Yearly Percentage Change
Warrant Arrests	7	11	31	40	29.0%
DUI Arrests	2	5	8	10	25.0%
Other Criminal Traffic Arrests	6	4	9	10	11.1%
Drug Arrests	3	10	16	31	93.8%

April 2005

Measuring "Catching the Bad Guys"

Source: Courtesy of John Gottschalk, Pierce County Sheriff's Dept., Washington.

What Works: Freedom to Analyze

The individual analyst benefits from freedom to be imaginative, to be free to pursue a trail, to explore, to dig deeper, to follow a path that only he or she might sense — one that may not even be articulated yet. One successful, experienced analyst avoids producing numerous standardized reports in order to have "down time," time where he can pursue the questions that have come about in the course of his work. He finds that it is in pursuing his questions that he discovers value — new information that can help his agency in unexpected ways. Just as patrol officers have free time to patrol proactively so should the analyst have free analytical patrol time to be proactive and able to respond to the unexpected.

What Works: The Generalist

Several subjects expressed concern about the development of training models and standardization of skills in the field of LE analysis They emphasized the need for a broad and diverse background to ensure a role for a variety of perspectives in the field. Being a generalist — someone who can do many things, like a "renaissance man" — was viewed as useful by some. Some felt that the artisan nature of analysis might be lost if the field were to be rigidly standardized.

One very experienced and accomplished analyst stated that forcing analysts into a narrow role would eliminate their power to make connections. He said that theories and methodologies risked circumscribing the analyst's view. Since "all intelligence analysis does is help us understand the world," limiting the scope of intellectual pursuits would hurt the profession. He encourages analysts to "take those intellectual flights of fancy," and that, despite the need for specific skill sets and knowledge in the post-9/11 world, the specific must be balanced by depth and breadth of knowledge and curiosity. He asserts that what we do "is magic."

What Works: The Specialist

Conversely, several individuals indicated that the need for specialists has emerged as technology advances. As the technical tools become more sophisticated, the demand for specialists with expertise grows.

What Works: The Artisan and the Engineer

Being able to visualize information to tell a story for others requires a special talent, according to some analysts. Writing gracefully is another talent. Some analysts emphasized the "art" aspect of the work. One expert, an engineer by training, indicated how important it is to think like an engineer. Rather than taking one's tools and applying them to a problem, he says it is much better to design tools especially suited to the problem at hand. Multi-disciplinary approaches are desirable – one individual pointed out that quality analysis depends on a "synthesis of disciplines."

What Works: Experience

A number of analysts assigned high value to their breadth of experience, especially those who had worked in a number of agencies and thought that "diverse multi-agency experience" made them better analysts. For them, this significant body of experience acted as a "good bank of ideas" and made the work process flow more intuitively. Experience provides a skill level that cannot be acquired by education or training.

> Experts can detect differences that novices cannot see, cannot even force themselves to see. Wine tasters can tell one grape from another and even one year of wine from another. To novices, wines are generic: they all taste the same. If you are just starting to drink wines, no matter how much attention you pay, how much you swirl the fluid around in your mouth, you don't get it. That's because "it" is not a fact (the Civil War began in 1861) or an insight (dividing one number into another is like subtracting it several times). You cannot learn just by being told or learn it all of a sudden. It takes lots of experience, and lots of variety in that experience, to notice differences. [64]

---

[64] Gary Klein, *Sources of Power: How People Make Decisions* (Cambridge, MA: The MIT Press, 2001), 157.

What Works: Persistence

A successful analyst will have a drive to understand what is going on, to provide a clear picture to others, and will find the best way to pursue the best problem-solving mode. Analysts are forced to be creative, to get the needed information, to gain credibility, to overcome organizational impediments. Along with persistence, the ability to discern "when enough is enough" is crucial. "Overdoing it is like NOT doing it at all."

What Works: People Skills

Developing an "inter-human connection" with everyone she meets was valued highly by one analyst and repeated, in other words, by a number of analysts. Developing personal relationships with officers was cited as extremely important. Gaining the trust of others so that one may eventually be able to benefit from their tacit knowledge is especially important.

Operating "as a value-added retail center," "becoming an "information brokerage," "training the consumer," all come under the umbrella of service orientation requiring higher order people skills. Talking to people in person, respecting them, knowing their names, task force physical co-location — all can contribute to improving people skills.

Certain analysts highly value that they "like people," describing this trait as conducive to making people comfortable with them, making others feel appreciated, which in turn helps them work on teams and get help in areas where they need help as well as earn appreciation themselves: "Relationships foster getting more of what you need: data sharing, technical support, and development."

WHAT WAS DISCOVERED

This is a Calling

In the project interviews many individuals expressed a desire to contribute to society; some analysts said they would even work for free because they found their work so meaningful. Some do, putting in extra unpaid hours to pursue an idea or complete an urgent project. A number of analysts called every day their high point, reporting that this work remains continuously attractive:

> "Every day is different."
> "This is a challenge every day."
> "I learn new things every day."

Choosing Good Analysts

Napoleon reportedly asked new officers on his staff one question: "Are you lucky?" Napoleon understood that luck in battle, although intangible, was a useful attribute.

A senior intelligence official used to ask his subordinates two questions about new analysts they wished to hire: "Do they think interesting thoughts? Do they write well?" This official believed that, with these two talents in hand, all else would follow with training and experience.[65] Law enforcement analysts at the local level do both very well, and much more.[66]

Next

Building upon individual strengths, the next level of success depends on relationships. Sharing of information and intelligence comes out of good personal relationships and thus these two topics — sharing and relationships — are grouped together in Chapter Four.

---

[65] Lowenthal, *Intelligence: From Secrets to Policy*, 2d ed. (Washington, DC: CQ Press, 2003), 91.

[66] The skills and abilities of individual analysts differ significantly; this text is referring to the analysts who are effective.

Chapter 4

WHAT WORKS: RELATIONSHIPS AND SHARING

At present, there are no standards related to information sharing and analysis.

—A Resource Guide to Law Enforcement, Corrections, and Forensic Technologies[67](U.S. DOJ, Office Of Justice Programs and Office Of Community Oriented Policing Services, May 2001), 54.

It's not what you know — it is all the people you know and the resources and assets they have that can help you.

—Anonymous interview subject.

In the terrible days following the 9/11 terrorist attacks, the entire nation became more conscious of the critical importance of sharing criminal justice and public safety information and intelligence among all branches and levels of government.

—Building Exchange Content Using the Global Justice XML Data Model: A User Guide for Practitioners and Developers[68] (U.S. DOJ, Office of Justice Programs: 17 June 2005), Introduction.

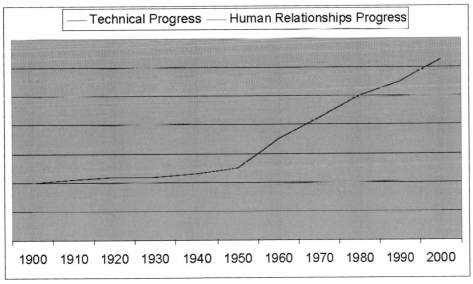

Paths of Technical and Human Relationships

Source: Chart concept adapted from William Glasser, *Choice Theory: A New Psychology of Personal Freedom* (New York: Harper Perennial, 1998), 9.

---

[67] Available online at *http://www.ncjrs.org/pdffiles1/nij/186822.pdf.*

[68] Available online at *http://it.ojp.gov/topic.jsp?topic_id=201.*

RELATIONSHIPS AND INFORMATION SHARING

The success of the intelligence cycle, however it is specifically defined and wherever it is applied, depends on the quality of relationships of those involved and their abilities to share information appropriately and effectively. The theme of relationships and sharing was emphasized in many of the interviews conducted for this project. Good relationships and the ability to share may seem like common sense topics that do not need articulation, but the gravity of failures to share and the infighting among those even now involved in homeland security, as well as within the law enforcement community, require examination and in-depth research beyond this project.

The preceding chart is adapted to illustrate this concept. It may over-dramatize the problem, but the problem itself is very real. Technology has progressed exponentially over the last century. *Applied* knowledge of how to improve human relationships (and thus facilitate sharing) has not. We may pour billions of dollars into systems that allow us to share information better. We may link everyone by the most innovative technological means such as hand-held instantaneous communication devices that can do myriad, unimagined things. All this can happen and is on the road to happening — but if we do not learn how to communicate better, how to relate better, how to share — none of the technology can affect some of our basic problems. Intelligence failures are often relationship failures. What *applied* knowledge works to improve this situation? This chapter highlights some of "what works" in relationships and sharing in law enforcement intelligence analysis.

A Very Brief History of Information Sharing

Sharing is foreign to national and criminal intelligence communities because of the classified nature of the work. Strict security is a doctrine. Even information regarding police 911 calls for service is restricted in some jurisdictions. Thus, new and urgent mandates to share involve real risk and, in many instances, changes in policies and ingrained practices.

> Generations of intelligence professionals have been trained in this distinction, the doctrine of disclosing information only to those who have a demonstrable "need to know," and the rigidities of the national security classification system. On the law enforcement side, it has long been recognized that confidentiality, protection of witnesses, and secrecy of grand jury information are essential to the successful investigation and prosecution of crimes. Thus, to both the law enforcement and foreign intelligence professions, proper security practices and strict limits on the sharing of information are second nature.[69]

---

[69] Statement by Eleanor Hill, Staff Director, Joint Inquiry Staff, *Hearing on the Intelligence Community's Response to Past Terrorist Attacks against the United States from February 1993 to September 2001*, 8 October 2002, 22.

The Risks of Sharing

Information may not equate to power in all situations, but information can be used in an unethical manner, and has been within law enforcement agencies and among them. There is a history of distrust and real reasons for it. Federal agencies that take the credit for investigations, when local law enforcement agency officers do the grunt work, add greater complexity to the mistrust between the levels of law enforcement. The risk of computer security problems adds to the concerns of many in law enforcement.

Police agencies have already experienced problems with low-level employees and even police officers selling information, from the janitor who funneled discarded drafts of the Jeffery Dahmer investigation to the news media to corrupt detectives providing case files to organized crime. Emerging technologies both exacerbate that age-old problem and add the new possibilities of direct interception and intervention without relying upon a human inside confederate.[70]

Sharing is facilitated by personal, face-to face relationships, even among rivals. Associating a face with a name, or being acquainted with someone who knows a face and can associate it with a name you know, makes a world of difference to many in law enforcement. Trust is a first-name-basis issue. Computers take away that relationship factor. Conversely, many people who never met in person are now connected through email and other computer-facilitated methods. They develop new types of relationships that facilitate unprecedented levels of information exchange.

ViCAP: An Example of a Systematic Failure in Information Sharing

The Violent Criminal Apprehension Program began in the FBI Academy in July 1985 with the goal of identifying possible violent serial criminals across the United States, based on crime pattern analysis through a voluntary computerized data collection system.[71] Since 1996, after discovering that at most eight percent of such crimes were entered into the database, with voluntary participation only by a few large urban areas, efforts have been made to modify the system to make it more user-friendly and increase participation. Even the name changed to "The New ViCAP."[72]

The concept is excellent — capture the data on unsolved homicides across the country and analyze it to uncover crime patterns that could conceivably result in early identification of serial killer behaviors and lead to earlier arrests. Old cases could be solved and investigators could pool resources. But because entry into the system is voluntary, the FBI must rely on the goodwill of each and every police agency to enter the data. This is inefficient and ineffective in uncovering threats. Considering all the movies and television

---

[70] Thomas J. Cowper and Michael E. Buerger, *Improving Our View of the World: Police and Augmented Reality Technology* (Washington, DC: Police Futures International/FBI, 2003), 55.

[71] Eric W. Witzig, "The New ViCAP," in *The FBI Law Enforcement Bulletin* 72, no. 6 (June 2003), 2.

[72] "The New ViCAP," 2-4.

shows about serial killers, the glamorous nature of CSI shows and other shows like it, and the public fascination with this topic — how could it be that we as compatriots cannot collect data on unsolved homicides in one place? There is no proactive way to analyze this heinous crime in our country. Agencies express concern that their information is out of their direct control if entered into this type of system.[73] At what cost do they exercise a freedom not to share?

The FBI has made distinct efforts to improve the situation by offering instructional and technological support along with the software. The software is free and ViCAP crime analysts are available to help agencies use the software for case management and case matching.[74] The failure in sharing is not the FBI's failure. It is a failure in the larger law enforcement system of our country, where for many civil-liberty and division-of-power reasons, we protect the privacy of serial killers and local priorities over national and even international concerns.

Why the Resistance?

Often, local law enforcement officials feel neglected by federal law enforcement agencies that have differing missions. Charles H. Ramsey, chief of the Washington, DC Metropolitan Police, said the federal agencies and local police departments had different missions and different needs after terrorists struck. "The F.B.I. is worrying about who might have done it, but what I care about is that there was an attack on a transit system and I have rush hour coming up," he said. "I don't need a threat analysis. I need to know what I can do proactively to strengthen the security of our transit system. Terrorism always starts as a local event," Chief Ramsey said. "We're the first responders."[75]

The police chiefs of up to fifteen cities have begun working together to create their own system of sharing terrorism-related information beyond the federal system. Frustrated by what they say is the slow and sometimes grudging way that federal officials share information about terrorist incidents, police chiefs from around the country are creating an informal network of more rapid communication.

Among the leaders is William J. Bratton, the Los Angeles police chief, who said in interviews Wednesday and Thursday that while the quality of information from the F.B.I. and the Department of Homeland Security was generally good, it often arrived far too late to be of any immediate value to local police departments. "The frustration is that intelligence gathering and sharing networks at the federal level are not working for local chiefs of police," Chief Bratton said. "We're used to things breaking very quickly and have to respond quickly. We don't have the luxury of waiting."[76]

---

[73] "The New ViCAP," 7.

[74] "The New ViCAP," 7.

[75] John M. Broder, "Police Chiefs Moving to Share Terror Data," *The New York Times,* 29 July 2005, online edition.

[76] "Police Chiefs Moving to Share Terror Data."

Recognition of the differing mission of local law enforcement and the federal ability to support it are necessary to build relationships. After all, each citizen is dependent on the local level of law enforcement for his or her primary safety needs. It is the local level of law enforcement we call when we have a crime problem or see suspicious behavior. Local law enforcement cannot wait for federal help. At the same time, federal recognition that local law enforcement generates data useful at the national level is overdue.

Systems/Initiatives for Intelligence Information Sharing

A number of systems exist for sharing intelligence. Although interview subjects place high value on relationships and sharing of information, very few mentioned these systems as things they currently value or see these systems as contributing to their successes.

Interstate information-sharing systems/initiatives in place or being developed at local, state, federal, and regional levels.[77]

CDU-Houston:
Community Defense Unit — Houston, Texas, Police Department

CISAnet:
Criminal Information Sharing Alliance Network
(Southwest Border States Anti-Drug Information System)

CLEAR-Chicago:
Citizen Law Enforcement Analysis and Reporting — Chicago area

COPLINK: various states

CriMNet-MN:
CriMNet — Minnesota

EFSIAC:
Emergency Fire Services Information and Analysis Center

EPIC:
El Paso Intelligence Center

ERN-Dallas:
Emergency Response Network — Dallas, Texas, FBI

HIDTA:
High Intensity Drug Trafficking Areas

---

[77]Read about most of these systems/initiatives at *http://it.ojp.gov/documents/intell_sharing_system_survey.pdf.*

JNET-PA:
Pennsylvania Justice Network

LEIU:
Law Enforcement Intelligence Unit

LEO:
Law Enforcement Online

LETS-AL:
Law Enforcement Tactical System — Alabama

MATRIX:
Multistate Anti-Terrorism Information Exchange

NCISP:
National Criminal Intelligence Sharing Plan

NLETS:
National Law Enforcement Telecommunication System

Project North Star:
Multi-agency law enforcement coordination coalition

RAID:
Real-time Analytical Intelligence Database

riss.net:
Regional Information Sharing Systems secure intranet

SIN-OK:
State Intelligence Network — Oklahoma

SPIN-CT:
Statewide Police Intelligence Network — Connecticut

TEW Group-Los Angeles Terrorism Early Warning Group — Los Angeles, California, area

ThreatNet-FL: web monitoring service

Source: Compiled by author.

An Example of Success in Sharing: CLEAR

The Chicago Police Department stands out as a center of innovation in law enforcement information sharing, not only among officers, but also with citizens.

CLEAR is an enterprise-wide vision of how anytime, anywhere access to centralized, relational data can empower intelligence-driven crime-fighting. While other police departments are struggling to integrate legacy data and applications, Chicago decided in the late 1990s that to have maximum impact, all policing intelligence should be accessible in one spot, with all tools leveraging and feeding that repository.

The CLEAR database, deployed in April 2000 and now topping 200GB, is the foundation for a growing set of integrated CLEAR applications used by all of the department's 13,600 officers and most of its 3,000 civilians, plus an exponentially expanding base of users outside the city limits. In fact, the Illinois state crime data system will be replaced by CLEAR, which will serve as the State Police's data repository. Typically, 1,200 concurrent users run more than 7,000 queries daily against data that include:

- Arrest reports
- Live cases' statuses
- Criminal activity by district, beat, street and address
- Rap sheets with aliases, nicknames and distinguishing physical marks
- Digital mug shots and fingerprints
- Seized property and evidence tracking
- Forensics reports
- Personnel data — number of arrests by each officer and many other performance metrics [78]

One of the individuals interviewed for this project, Dr. Michael Maltz, worked in Chicago when it originally started using ICAM (Information Collection for Automated Mapping), a crime mapping and analysis application, and was a lead developer of the ICAM project, the forerunner of CLEAR.[79] Maltz says that the improved, realistic visualization of data, such as the use of ortho photography and animation, will facilitate "knowing the whole of ... everything that has occurred." The map becomes the collective memory of the department, since it captures the information from every shift and every officer. Animation allows officers and analysts to look over various time intervals, daily, weekly, seasonally — whatever criteria are set, and, by factoring in other information, such as how much light or precipitation there was at a given time, it can help everyone become a crime analyst and conduct analysis in near-real-time. This is a futuristic, but realistic, hope.

---

[78] Richard Pastore, "Taking IT to the Street: How the Chicago Police Department Used Technology to Fight Crime and Become the First Grand CIO Enterprise Value Award Winner," *CIO Magazine*, 15 February 2004. URL *http://www.cio.com/archive/021504/grand.html.*

[79] The city of Chicago uses Appreciative Inquiry on a wide-scale; *Imagine Chicago* was launched in 1992 and is an ongoing project — see www.imaginechicago.org.; *http://www.abtassociates.com/reports/icamprog.pdf.*

An Example of Success in Sharing: The Sniper Investigation

A Police Executive Research Forum report highlights what worked in the sniper investigation, the efforts to end the killing spree in the Washington, DC area, in October 2002:

> Law enforcement needs systems that allow multiple agencies in complex investigations to exchange and analyze information. In the sniper case, talented and dedicated information systems specialists and crime analysts were forced to patch together portions of different systems to create an information analysis system that eventually provided the investigation with robust intelligence capabilities.[80]

> *Of all the keys to success during this investigation, the one mentioned — indeed, strongly emphasized — by every executive was the importance of pre-existing relationships.*[81]

The Importance of "Weak" Social Ties

The concept of the power of "weak" social ties applies to the success of LE intelligence analysis. The more people you know outside of your agency, the more access you have to information. The weak ties concept is described in this passage:

> Weak ties play a crucial role in our ability to communicate with the outside world. Often our close friends can offer us little help in getting a job. They move in the same circles we do and are inevitably exposed to the same information. To get new information, we have to activate our weak ties. Indeed, managerial workers are more likely to hear about a job opening through weak ties (27.8 percent of the cases) than through strong ties (16.7 percent). The weak ties, or acquaintances, are our bridge to the outside world, since by frequenting different places they obtain their information from sources other than our immediate friends.[82]

In brief, it is not only who you know, but as important, and maybe even more important, it is *who they know*. This concept echoes the message in Surowiecki's, *The wisdom of Crowds*.

A Working Model for "Weak" Social Ties

The importance of human relationships was emphasized over and over again in the interviews conducted for this study. A model of effective relationship is that of the National Drug Intelligence Center's Field Program Specialist. In New York Sate, Eddie Beach holds this title and as part of his networking job tasks has face-to-face contact with

---

[80] See *Managing a Multijurisdictional Case: Indentifying the Lessons Learned from The Sniper Investigation,* http://www.ojp.usdoj.gov/BJA/pubs/SniperRpt.pdf.

[81] *Managing a Multijurisdictional Case,* 25.

[82] A.L. Barabasi, *Linked: How Everything is Connected to Everything Else and What It Means for Business, Science, and Everyday Life* (New York: Penguin Group, 2003), 43.

officers, analysts, and health care agencies. If there is an issue, problem or question related to drugs, Eddie is always there to help or inform. For example, when there was a rash of heroin deaths in western New York, he was able to keep everyone informed and help investigators link suspected overdose deaths to a particularly dangerous batch of heroin. He can bring information to Washington or vice versa. Real people interfacing with real people is a highly effective practice, according to many of the subjects interviewed. Developing creative ways to leverage this concept may help improve intelligence analysis and sharing capacities.

Individuals acting as information "hubs" in a network could link many people, effectively overcoming the challenge of sharing. To design such a system proactively and wisely may be a step toward facilitating information sharing. Analysts, being effective "knowledge brokers" in law enforcement, could be effective hubs.

U.S. Coast Guard Operations Specialist Robert Ziehm also emphasized the importance of face-to-face relationships. As an experienced law enforcement investigator, now retired from that work, he says that people are more apt to open up in person and provide information beyond shared written reports.

A Second Working Model of "Weak Tie" Social Relationships

One of the most significant contributions to the development of law enforcement analyst relationships comes from the listserv LEAnalyst,[83] a clear example of the benefit of weak-tie social relationships. LEAnalyst links over 1,500 law enforcement analysts, experts, and curious bystanders in an open network of sharing. Post a question on LEAnalyst and a leading world expert may help you with an answer. You probably don't know these people personally, but they "know" you because you belong to LEAnalyst. They recognize your name, or know you are an analyst.

Other informal, but restricted, listservs created by field officers in intelligence are emerging in ad hoc fashion, connecting voluntary members in the military, national security, and federal, state, and local law enforcement. This trend may be an indicator that, although these structures at times experience difficulties in blending resources and missions, the ground-level workers have found ways around the obstacles of bureaucracy and power struggles to do the work they value.

---

[83] See *http://www.leanalyst.info/*. "The primary purpose of the LEAnalyst mailing list is to provide a place where law enforcement employees (sworn and non-sworn), academia (instructors and students), and businesses providing products or services to the law enforcement community can meet and exchange information, methods, and ideas regarding the analysis of crime. We allow other law enforcement-related postings in order to encourage a free flow of information regarding all manner of law enforcement concerns." The listserv is maintained by Ray Sanford and Sal Perri as a free service.

WHAT WORKS

Global Justice XML Data Model

Several subjects said that the Global Justice XML (Extensible Markup Language) Data Model (Global JXDM) was an important milestone toward more effective information-sharing systems.

Developed by the U.S. Department of Justice's Global Justice Information Sharing Initiative, the Global JXDM is a common vocabulary that is understood system to system, enabling access from multiple sources and reuse in multiple applications, allowing justice and public safety communities to effectively exchange information at all levels.[84]

Paul Wormeli, a participant in this research project, gives an example of how the Global Justice XML Data Model "works":

The whole idea of the GJXDM is to make it very easy to share information between law enforcement agencies... also to expedite judicial processing by sending incident and arrest data easily to prosecutors and courts in near real time so that the prosecution and judicial proceedings are expedited. Information sharing in the dynamic and highly mobile world of crystal meth is a big key to law enforcement success through collaboration in cases involving multiple people, locations, and tactics, irrespective of city boundaries. The more they share, the greater the likelihood of successful case clearance and effective prosecution.

As states implement prescription drug monitoring programs and develop the system for exchanging information across state lines, drug dealers specializing in the illicit traffic of controlled substances will have a hard time hiding from law enforcement agencies.[85]

Broad Networking

In LE analysis, success depends upon a high degree of networking, and crime analysis itself has helped the growth of law enforcement networking. "Nobody seems to network like analysts do. We ask each other for help in unprecedented ways." One analyst points out that officers do not seem to ask one another *how* to do their jobs, which runs counter to the common activity of most analysts in LE — asking peers instead for help, [and for] guidance in improving one's skills, techniques, or access to resources.

One national-level analyst said that the interview (in the context of his work in a national policing agency and the author's work as a local-level law enforcement analyst)

---

[84] See URL *http://it.ojp.gov/topic.jsp?topic_id=201.*

[85] "Catching Drug Dealers: GJXDM To The Rescue," *The Rockley Bulletin at http://www.rockleybulletin.com/technologycorner_comments.php?id=99_0_6_0_C.*

demonstrated the valued ability to reach across communities and disciplines and find that the work we do is fundamentally the same trade.

Cooperative Internal Relationships

Several analysts emphasized the importance of nurturing and maintaining cooperative relationships among patrol personnel, investigators, and administrators. Learning about the type of work done by others can help analysts in a number of ways. They may discover information they did not know existed, they may discover an analytical product needed, and they will gain respect and trust.

Relationships with IT Staff

Developing relationships with IT (Information Technology) people is an approach that works; the relationship benefits from any opportunity analysts have to show the IT personnel what analysts do, thus making clear how IT specialists can contribute to the team effort. Often, the tools the analysts need and use put stress on the IT office and the agency's IT system itself — the specialist's responsibility is to make sure the system performs and the analyst's work does not unduly slow the operation of systems. Without a personal relationship between the two specialties, the likelihood of system overload can cause conflict unanticipated by an isolated analyst. Developing good relationships with IT support staff by showing them why their help is needed can facilitate improvement in getting analytical tools to work optimally in an agency.

Interview subjects made it clear that IT people often do not understand the greater importance of interpreting data — being able to engage in intensive manipulation of data — compared to the lesser value inherent in information-storage capabilities: "It is a specific name that solves a case, not just data."

Teamwork

Law enforcement intelligence analysts were more likely than traditional crime analysts to emphasize the importance of teamwork, probably because they work more often on investigative teams. One senior analyst saw the intelligence analyst's role as complementary to the investigator's role, reporting that investigations without intelligence-oriented analytical support are "deficient." Intelligence analysts, according to him, provide intangible but valuable insights. Several law enforcement intelligence analysts reported in interviews that they enjoyed the camaraderie of law enforcement teamwork, the sense of working together toward the common goal of public safety.

Developing intelligence in a team setting, face-to-face, is recommended by Coast Guard intelligence analyst Robert Ziehm. Using charts to visualize who/what/when/where in time and space, to pool resources to find better opportunities to detect illegal border activity, and to design operations around analytical findings are cost-effective keys to greater success.

Relationships with the Community[86]

Local-level law enforcement intelligence analysts are in a different position from most other analysts in that they can and do, in some jurisdictions, attend community meetings and other types of problem-solving efforts where the stakeholders are working together. Some of the analysts are involved with providing citizens with data, surveying citizens, and working on task forces consisting of agencies outside of law enforcement. They may attend block club meetings, recommend responses that include warning citizens of a current crime series, or help local government assess specific issues related to police data.

> Our interest is in solving real world problems with research and analytical techniques. As such our position is that research must be directed in practical ways. As good ethnographic researchers (and good street cops) will tell you, some of the very best ideas for problem-solving and research can come from those being researched! It can come from kids in the schoolyard, local taxi drivers, street hookers, beat cops and residents in the apartment.[87]

INFORMATION SHARING AND HOMELAND SECURITY

It should be clear by now that there is not one law enforcement community, but many.

The tragedy of 9/11 forced the Intelligence and Law Enforcement Communities to reevaluate their information sharing practices. The nature of the threat today is such that there cannot be a dichotomy between the two communities with regard to information; all available relevant intelligence information from one community must be shared with the other in order to generate the most comprehensive analysis of the potential threat to our homeland.[88]

National Criminal Intelligence Sharing Plan[89]

Developed by the Global Intelligence Working Group (GIWG) and endorsed by U.S. Attorney General John Ashcroft, U.S. Department of Justice (DOJ), the *National Criminal Intelligence Sharing Plan* (NCISP) is a formal intelligence sharing initiative that addresses the security and intelligence needs recognized after the tragic events of 11 September 2001. It describes a nationwide communications capability that will link together all levels of law enforcement personnel, including officers on the streets, intelligence analysts, unit commanders, and police executives for the purpose of sharing critical data.[90]

---

[86] See Thomas Rich, *Crime Analysis and Mapping by Community Organizations in Hartford, Connecticut* at *http://www.ncjrs.org/pdffiles1/nij/185333.pdf* for an example of empowering neighborhoods to utilize crime mapping for problem solving.

[87] Greg Saville and Chuck Genre, draft manuscript *Understanding Neighborhood Problems,* University of New Haven.

[88] Christopher C. Thornlow, *Fusing Intelligence with Law Enforcement Information: An Analytic Imperative* (Monterey, CA: Naval Postgraduate School Thesis, March 2005), 49.

[89] See document at *http://it.ojp.gov/documents/200507_ncisp.pdf.*

[90] See NCISP home page at *http://it.ojp.gov/topic.jsp?topic_id=93.*

The National Criminal Intelligence Sharing Plan was developed to address:

1. **Lack of communication and information sharing** — specifically, lack of a centralized analysis and dissemination function, either at the state or federal level; lack of intelligence from federal agencies; and state statutory requirements that present hurdles to sharing information.

2. **Technology issues** — specifically, lack of equipment to facilitate a national intelligence data system, lack of interconnectibility of law enforcement and other databases (that is, immigration services), limited fiscal resources, lack of technological infrastructure throughout the state, and lack of uniformity between computer systems.

3. **Lack of intelligence standards and policies** — specifically, lack of common standards for collection, retention, and dissemination of intelligence data; a need for increased local training on legal standards for collection, storage, and purging of data; access to classified data; and lack of standards for determining when to disseminate intelligence to federal agencies.

4. **Lack of intelligence analysis** — specifically, lack of compatible analytical software and lack of analytical support, personnel, equipment, and training.

5. **Poor working relationships** — specifically, unwillingness of law enforcement agencies to provide information due to parochial interests and a culture within the federal system that does not foster sharing of information or trust between agencies.[91]

An irony of the NCISP is that neither the International Association of Crime Analysts, nor active crime analysts at the local level of LE, were involved in its conception or development. Good relationships must be fostered between the analyst associations as well. Law enforcement leaders must also recognize that what crime analysts contribute is also *intelligence*.

Next

Law enforcement has partnered with academia to develop several processes that significantly change the way police work is done compared with traditional policing — processes in which intelligence analysis is a core component. Other innovative policing processes have emerged due to change in police management priorities. Many subjects cited these processes as important in their work, in bringing success and in preparing for the future. The processes build upon individual analytical strengths, relationships, and sharing. Chapter Five highlights the processes.

---

[91] NCISP home page, 3.

Chapter 5

WHAT WORKS: PROCESSES

Traditional policing has a narrow focus, wanting to latch onto a guy rather than ask, *What are the possibilities? Where can you get the best bang for your buck?* [92]

In the Intelligence Community, the intelligence cycle is the process central to intelligence analysis. In law enforcement intelligence circles, the emergence of several different approaches to policing, rooted in multi-disciplinary academic research, has resulted in the institutionalization of numerous processes that have analysis as a central feature. Analysts and expert observers value these processes. For law enforcement intelligence analysts, the processes provide an avenue for increasing the value of their role. For contributing experts, the processes provide a venue for innovative research and analysis methods to bear additional fruit in the real world.

INTELLIGENCE-LED POLICING

As noted in Chapter One, the United Kingdom and the United States, along with Australia, Canada, and several other countries, are moving toward a new form of policing called *intelligence-led policing.*

Intelligence-led policing is more than just a catchy phrase. We are at a critical time when we should be collaborating to jointly develop a clear definition of what it means. Based on current literature, I believe intelligence-led policing is about being able to:

describe, understand and map criminality and the criminal business process; make informed choices and decisions;
engage the most appropriate tactics;
allow the targeting of resources;
disrupt prolific criminals; and
articulate a case to the public and in court.

To retool, to implement intelligence-led policing, a strategically directed police effort must be based on analysis. It is only by focusing on intelligence work performed by a specialized workforce that we can be released from the predominately reactive cycle of policing.

Source: R. Ban, "The Dynamics of Retooling and Staffing: Excellence and Innovation in Police Management," Unpublished Candian Police College manuscript, 2003, 3.

---

[92] Quote from interview.

Intelligence-led policing represents a paradigm shift. In the United Kingdom, its implementation rests on a formal business model of policing, or corporacy.

The Concept of Corporacy

Looking at the business of policing as one might look at a corporation is radically different from the way policing has been done in the past.

> Good corporacy occurs when the leaders in a sector have a common understanding of the activities, behaviours and results that create value for the customers of the sector; it is tied to the concept of "best value." Increasingly, the funding of public sector services such as policing and health care is contingent upon demonstrating to government funding and oversight agencies that the service performed with public funds fulfills government objectives and meets standards set. Thus it is no longer good enough just to patrol neighbourhoods and catch criminals; a police service must show that its efforts are, for example, reducing crime levels and making the community feel safer, and furthermore that we are doing so in the most efficient and effective possible way. Many police services have started the challenging process of retooling to make performance measurement an integral part of operations.[93]

In the corporacy view of intelligence-led policing, it is not enough for officers to be successful by arresting people, by investigations that close down major drug suppliers, or by answering 911 calls within minutes. The new measurement of success will examine the impact of the work. Does crime go down? Did the drug supply dry up? Did we eliminate demands for service (911 calls) by solving underlying chronic problems instead of simply

---

[93] "The Dynamics of Retooling and Staffing," 7.

reacting to them individually? If not, can we think of other ways to do business that will get the results our customers, the citizens, want?

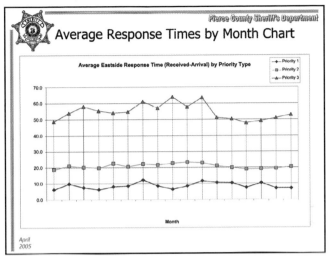

What does this chart mean for the community?

Source: John Gottschalk, crime analyst, Pierce County Sheriff's Office, Washington State.

A number of interview subjects are excited about intelligence-led policing, but no one had examples of success that they attributed specifically to this process. Their excitement revolves around the fact that *analysis* is the driver of intelligence-led policing. Intelligence-led policing cannot exist without analysis.

COMPSTAT

The COMPSTAT process has been adapted by numerous law enforcement agencies. One of the author's sources, Joseph Concannon, calls himself "one of the babies of COMPSTAT." A retired captain of the NYPD, where COMPSTAT originated, he notes that "the beauty of COMPSTAT is communication." The maps and statistics provide visualization of crime and other policing problems to a whole group, facilitating discussion so that decisions can be made, and workers held accountable. "Every aspect of COMPSTAT involves measurement," according to Concannon.

At the turn of the twenty-first century, a new engine of police progressivism may have arisen. Characterized as a new crime-control program, COMPSTAT combines all of the major prescriptions offered by contemporary organizational development experts with the latest geographic information systems technology. It also re-engineers police management by holding command staff directly accountable for crime levels in their beats. Furthermore, it uses sophisticated computer maps and crime statistics to facilitate timely and targeted responses to crime problems.[94]

Crime analysts value the concept of COMPSTAT because it has created a demand for and thereby elevated the value of their work. Agencies that measure their crimes, look for emerging and chronic problems, and seek to develop strategies to address them, need crime analysis. However, crime analysis should NOT be confused with COMPSTAT. The technical ability to make maps and generate statistics is not analysis. Often agencies confuse the two — generating charts and maps rather than analyzing the data to see what they mean.

COMMUNITY POLICING

The Office of Community-Oriented Policing Services (COPS)[95] offers a definition of Community Policing:

> Community policing focuses on crime and social disorder through the delivery of police services that includes aspects of traditional law enforcement, as well as prevention, problem-solving, community engagement, and partnerships. The community policing model balances reactive responses to calls for service with proactive problem-solving centered on the causes of crime and disorder. Community policing requires police and citizens to join together as partners in the course of both identifying and effectively addressing these issues. [96]

The COPS Office funded the project that resulted in the publication of *Law Enforcement Intelligence: A Guide for State, Local, and Tribal Law Enforcement Agencies.*[97] The principal investigator was David L. Carter, from the School of Criminal Justice, Michigan State University. Chapter Four of the book is called "Intelligence-Led Policing (ILP): The Integration of Community Policing and Law Enforcement Intelligence." Carter asserts in this chapter that community policing and intelligence-led policing bear similarities:

---

[94]The Police Foundation report, *Compstat in Practice: An In-Depth Analysis of Three Cities at http://www.policefoundation.org/pdf/compstatinpractice.pdf.*

[95] "The COPS Office was created as a result of the Violent Crime Control and Law Enforcement Act of 1994. As a component of the Justice Department, the mission of the COPS Office is to advance community policing in jurisdictions of all sizes across the country. Community policing represents a shift from more traditional law enforcement in that it focuses on prevention of crime and the fear of crime on a very local basis. Community policing puts law enforcement professionals on the streets and assigns them a beat, so they can build mutually beneficial relationships with the people they serve. By earning the trust of the members of their communities and making those individuals stakeholders in their own safety, community policing makes law enforcement safer and more efficient, and makes America safer." from *http://www.cops.usdoj.gov/default.asp?Item=35 retrieved 7/21/05.*

[96] See URL *http://www.cops.usdoj.gov/default.asp?Item=36.*

[97] U.S. Department of Justice, *Law Enforcement Intelligence: A Guide for State, Local and Tribal Law Enforcement Agencies* (Washington, DC: Office of Community Oriented Policing Services, 2004).

Both community policing and ILP rely on:

- Information Management
  - ❏ Community policing — Information gained from citizens helps define the parameters of community problems.
  - ❏ ILP — Information input is the essential ingredient for intelligence analysis.
- Two-way Communications with the Public
  - ❏ Community policing — Information is sought from the public about offenders. Communicating critical information to the public aids in crime prevention and fear reduction.
  - ❏ ILP — Communications from the public can provide valuable information for the intelligence cycle. When threats are defined with specific information, communicating critical information to citizens may help prevent a terrorist attack and, like community policing, will reduce fear.
- Scientific Data Analysis
  - ❏ Community policing — Crime analysis is a critical ingredient in the COMPSTAT process.
  - ❏ ILP — Intelligence analysis is the critical ingredient for threat management.
- Problem Solving
  - ❏ Community policing — Problem solving is used to reconcile community conditions that are precursors to crime and disorder.
  - ❏ ILP — The same process is used for intelligence to reconcile factors related to vulnerable targets and trafficking of illegal commodities.[98]

Several analysts stressed the importance of the community in the work of analysis: "If you want to solve crimes, your connection with the community is invaluable." Participant Greg Saville takes this concept even further. He believes that unless the stakeholders of the community are involved in the problem-solving process, it is doomed to failure because it is irrelevant. Unless the people with the problems are involved in the process, he is pessimistic that real change can occur. "Crime analysis is a tool to be used to do something else; it can help build capacities in neighborhoods to prevent crime."

Community policing has been implemented in many law enforcement agencies in varying degrees, depending on the interpretation of the concept and available resources. As of 30 June 2000,

- Two-thirds of all local police departments and 62% of sheriffs' offices had full-time sworn personnel engaged in community policing activities.

---

[98] Law Enforcement Intelligence: A Guide for State, Local, and Tribal Law Enforcement Agencies, 41-42.

- Local police departments had an estimated 102,598 full-time sworn personnel serving as community policing officers or otherwise regularly engaged in community policing activities, and sheriffs' offices had 16,545 full-time sworn so assigned. [99]

PROBLEM-ORIENTED POLICING

The concept of problem-oriented policing (POP) was articulated by Herman Goldstein in his seminal *Problem-Oriented Policing* in 1990.[100] It is a central strategy advocated by the COPS Office.

> Problem-oriented policing is an approach to policing in which discrete pieces of police business (each consisting of a cluster of similar incidents, whether crime or acts of disorder, that the police are expected to handle) are subject to *microscopic examination* (drawing on the especially honed skills of crime analysts and the accumulated experience of operating field personnel) in hopes that what is freshly learned about each problem will lead to discovering a *new and more effective strategy* for dealing with it. Problem-oriented policing places a high value on new responses that are *preventive* in nature, that are *not dependent on the use of the criminal justice system*, and that *engage other public agencies, the community and the private sector* when their involvement has the potential for significantly contributing to the reduction of the problem. Problem-oriented policing carries a commitment to *implementing* the new strategy, rigorously *evaluating its effectiveness,* and, subsequently, *reporting* the *results* in ways that will benefit other police agencies and that will ultimately contribute to building a body of know-ledge that supports the further professionalization of the police. [101]

POP is a process residing within other processes. ILP and Community Policing both stress its importance. It is well-documented through numerous case studies that exemplify how to implement the process.

---

[99] See *http://www.ojp.usdoj.gov/bjs/sandlle.htm.*

[100] Herman Goldstein, *Problem-Oriented Policing* (New York: McGraw-Hill, 1990).

[101] From the COPS Center for Problem Oriented Policing at URL *http://www.popcenter.org/about-whatisPOP.htm.*

Some Comparisons between Problem-Oriented Policing and Community Policing Principles[102]

Principle	Problem-Oriented Policing	Community Policing
Primary emphasis	Substantive social problems within police mandate	Engaging the community in the policing process
When police and community collaborate	Determined on a problem-by-problem basis	Always or nearly always encouraged, but less important than community collaboration
Emphasis on problem analysis	Highest priority given to thorough analysis	Preference for collaborative responses with community
Preference for responses	Strong preference that alternatives to criminal law enforcement be explored	Emphasizes strong role for police
Role for police in organizing and mobilizing community	Advocated only if warranted within the context of the specific problem being addressed	Essential
Importance of geographic decentralization of police and continuity of officer assignment to community	Preferred, but not essential. Strongly encourages input from community while preserving ultimate decision-making authority for police	Emphasizes sharing decision-making authority with community
Degree to which police share decisionmaking authority with community	Emphasizes intellectual and analytical skills	Emphasizes interpersonal skills
Emphasis on officers' skills; view of the role or mandate of police	Encourages broad, but not unlimited role for police, stresses limited capacities of police and guards against creating unrealistic expectations of police	Encourages expansive role for police to achieve ambitious social objectives

Source: author.

---

[102] Michael S. Scott, "Relating Problem-Oriented Policing to Other Movements in Police Reform and Crime Prevention," *Problem-Oriented Policing: Reflections on the First 20 Years* (USDOJ:COPS, 2000), 99. Available online at *http://www.cops.usdoj.gov/default.asp?Item=311*. "This monograph assesses the current state of problem-oriented policing, revisits the fundamental principles of Herman Goldstein's POP framework, and reports on the successes and distortions in implementing POP over the last 20 years. It is an invaluable resource for those seeking to better understand this fundamental element of policing."

EXAMPLES OF SUCCESS: DOCUMENTED PROCESSES THAT WORK

The book, *Become a Problem-Solving Crime Analyst in 55 Small Steps*,[103] by Ron V. Clarke and John Eck, both of whom participated in this project, leads readers through the problem-solving process as experienced by working analysts. The emphasis is on problem solving, rather than just apprehending "the bad guys." A U.S. version, *Crime Analysis for Problem Solvers in 60 Small Steps*, funded by the COPS Office, is also available.[104] The book emphasizes social science research skills to address crime problems.

Problem-Oriented Policing Guides[105] at the U.S. Department of Justice Office of Community Oriented Policing Services are also available online. They cover interesting topics that affect a wide swath of people in any community.

- Gun Violence among Serious Offenders
- Illicit Sexual Activity in Public Places
- Identity Theft
- Loud Car Stereos
- Misuse and Abuse of 911
- Panhandling
- Prescription Fraud
- Rave Parties

---

[103] Available online at *http://www.jdi.ucl.ac.uk/publications/manual/crime_manual_content.php*.
[104] See *http://www.popcenter.org/Library/Recommended Readings/60Steps.pdf*.
[105] *http://www.cops.usdoj.gov/Default.asp?Item=248*.

The Center for Problem-Oriented Policing [106] endorses the steps outlined in the following chart.

25 Techniques for Problem-Oriented Policing

Increase efforts	Increase risks	Reduce rewards	Reduce provocations	Remove excuses
1. Harden targets Immobilizers in cars; Anti-robbery screens	6. Extend guardianship Cocooning; Neighborhood watch	11. Conceal targets Gender-neutral phone lists; Off-street parking	16. Reduce frustration and stress Efficient queuing; Soothing lighting	21. Set rules Rental agreements; Hotel registration
2. Control access to facilities Alley-gating; Entry phones	7. Assist natural surveillance Improved street lighting; Neighborhood watch hotlines	12. Remove targets Removable car parts; Pre-paid public phone cards	17. Avoid disputes Fixed cab fares; Reduce crowding in bars	22. Post instructions "No parking"; "Private property"
3. Screen exits Tickets needed; Electronic tags for libraries	8. Reduce Anonymity Taxi-driver IDs; "How's my driving?" signs	13. Identify property Property marking; Vehicle licensing	18. Reduce emotional arousal Controls on violent porn; Prohibit pedophiles working with children	23. Alert conscience Roadside speed display signs; "Shoplifting is stealing"
4. Deflect offenders Street closures; Separate toilets for women	9. Use place managers Train employees to prevent crime; Support whistle-blowers	14. disrupt markets Check pawn brokers; License street vendors	19. Neutralize peer pressure "Idiots drink and drive"; "It's ok to say no"	24. Assist compliance Litter bans; Public lavatories
5. Control tools/weapons Toughened beer glasses; Photos on credit cards	10. Strengthen formal surveillance Speed cameras; Closed-circuit TV in town centers	15. Deny benefits Ink merchandise tags; Graffiti cleaning	20. Discourage imitation Rapid vandalism repair; V-chips in TVs	25. Control drugs/alcohol Breathalyzers in bars; Alcohol-free events

Source: Modified from D.B Cornish and R.V Clarke, "Opportunities, Precipitators and Criminal Decisions: Reply to Wortley's Critique of Situational Crime Prevention," in M. Smith and D.B. Cornish, eds., *Theory for Situational Crime Prevention.* Crime Prevention Studies, Vol. 16. (Monsey, NY: Criminal Justice Press, 2003), 90. See URL: *http://www.popcenter.org/25techniques.htm.*

---

[106] The Center also offers innovative interactive learning modules such as this example: Interactive Learning Module: Street Prostitution at *http://www.popcenter.org/learning.htm.*

Problem Analysis Triangle

The Problem Analysis Triangle was derived from the routine activity approach to explaining how and why crime occurs. This theory argues that when a crime occurs, three things happen at the same time and in the same space:

- a suitable target is available;
- there is the lack of a suitable guardian to prevent the crime from happening at that time and place;
- a motivated offender is present.

Problem-Analysis Triangle

Source: Graphic and text modified from the Community-Oriented Policing Center for Problem-Oriented Policing at *http:// www.popcenter.org/about-triangle.htm.*

As an example of how this paradigm generates tools and approaches in crime prevention, the UK's Home Office has a number of "Crime Reduction Toolkits" available online.[107] The Introduction to the Toolkit "Using Intelligence and Information" follows.

The use of sound, quality information and intelligence processes are essential to identifying and limiting the activities of those committing crime and disorder and tackling the problems which adversely affect community safety and quality of life.

---

[107] *http://www.crimereduction.gov.uk/toolkits/.*

Information gathering and intelligence lies at the heart of business planning by taking into account local and national government objectives, required levels of performance and value-for-money principles. The core ingredient in successful business planning is information and understanding on five key issues:

- An accurate picture of the business;
- What is actually happening on the "ground";
- The nature and extent of the problems;
- The trends;
- Where the main threats lie.

Confident and effective information exchange is the key to multi-agency cooperation. It relies on good relations between partners, and on mutual trust. The effectiveness of information exchange arrangements is a reflection of the effectiveness of the partnership as a whole.

There are many good reasons for exchanging information:

- Combining information resources creates a more accurate picture of what is going on in the local area.
- Better problem analysis and sound decision-making flow from the possession of good intelligence and accurate data.
- Key decisions on how to invest and allocate resources will be more effective and easily justified, if they are information-based.
- A combination of sound intelligence and effective interventions targeting crime and disorder in an area reduces victimization, removes fear and benefits all sectors of a community, except the criminal.

The case for sharing information to achieve these objectives is indisputable. This toolkit must be considered in conjunction with the other crime reduction toolkits before developing particular strategies and action plans. In the words of the UK's Home Office:

> This Toolkit covers two main areas: analytical techniques and products for effective intelligence/information; and processes for effective information sharing. Work will be carried out to improve the content and ensure focus on the needs of crime and disorder reduction partnerships. The effectiveness of the toolkits relies on your help. We very much welcome contributions and advice on how to improve the content and approach. We also particularly welcome examples of good local practice which has been shown to work. [108]

---

[108]See *http://www.crimereduction.gov.uk/toolkits/ui01.htm.*

THE GROWTH OF CRIME ANALYSIS BASED ON NEW PROCESSES

Noah Fritz, current president of the International Association of Crime Analysts, describes how the growth of crime analysis is based on a number of factors, including new processes.

The profession of crime analysis has grown exponentially over the past 15 years. One of the most interesting questions regarding crime mapping and analysis remains: Why has this current growth and interest in crime analysis occurred at this point in time? The modern popularity of Crime Analysis can be largely explained by five primary reasons: (1) cost and popularity of computers have dramatically dropped and increased, respectively, both in society and within law enforcement since 1990; (2) the conceptualization of community and problem-oriented policing and subsequent publications by academicians has legitimized the role of research and analysis in policing; (3) the federal government has financially backed this initiative with personnel and technology grants; (4) chiefs of police and sheriffs have embraced the concept and practice of community policing, culminating in the success of COMPSTAT and ICAM (Information Collection for Automated Mapping) and the claims associated with this effort have pointed to crime rate reductions within jurisdictions utilizing this new style of policing; finally, (5) professional standards have been identified and established within law enforcement; specifically, the Commission for the Accreditation of Law Enforcement Agencies (CALEA) has provided support of crime analysis. [109]

The processes of Intelligence-Led Policing, COMPSTAT, Community Policing, and Problem-Oriented Policing have led to the growth of analysis and the movement toward developing a *small* but growing cadre of very skilled analysts in law enforcement at the local-level. The impact of these processes on the state or federal levels of law enforcement is unclear. The intelligence analysts at state and federal levels, along with a few local-level analysts, cited the ILP process as very important even though it is in its early stages. Many of the local law enforcement analysts cited the other processes as valuable, especially Problem-Oriented Policing. Both ILP and POP place emphasis on and rely on analysis — it is central to each approach.

The Limitations of the Processes

Despite the emergence of the processes explained in this chapter and the new recognition of the importance of analysis, the growth of professional intelligence analysis in law enforcement is inadequate to meet the demands for analysis implicit in achieving widespread success with any of the policing approaches.

---

[109] Noah Fritz, "The Growth of a Profession: A Research Means to a Public Safety End," *Advanced Crime Mapping Topics*, National Law Enforcement & Corrections Technology Center, 2002, 3. This list builds on ideas set forth in John E. Eck and David Weisburd, eds., *Crime and Place* (Monsey, NY: Criminal Justice Press, 1995), and David Weisburd and Tom MeEwen, *Crime Mapping and Crime Prevention* (Monsey, NY: Criminal Justice Press, 1997), Introduction.

Too often, law enforcement agencies and policymakers praise the new ways of doing business yet do not provide the resources to engage in the processes, including resources for analysis and analysts. Without a redirection of resources there will be no change. Law enforcement systems themselves resist change, sometimes employing one of the processes in name only, and going on about business as usual. Individual analysts and officers do have some ability to implement the processes on their own, but unless the whole organization changes its mental model of what law enforcement is and what it is supposed to do — the philosophy of agencies themselves — the value of these processes is limited. Their successful application requires a commitment to broad change.

Intelligence-led policing is in its very early stages. Obstacles to its implementation are the same as obstacles to the integration of the many agencies involved in homeland security. The need for classification of materials, the need for trust and relationships, the need for a common definition of "intelligence" — these are but a few of the things that stand in the way of a system of law enforcement that bases its decisions and deployment of resources on timely, accurate, relevant, analyzed information.

COMPSTAT has its critics. It lacks a strategic focus in its admirable mission to deal with immediate problems as swiftly as possible. Accountability for police managers is both its strength and its weakness. Having measurable goals, being rewarded with recognition for success and being helped with one's own professional area by receiving additional resources when needed are positive outcomes for those managers working in a COMPSTAT model. Conversely, problems with morale are the result of employing accountability meetings that humiliate individuals and instill excess fear and unreasonable blame.

Community Policing, at its best, empowers all stakeholders with information to address crime and disorder problems. Ideally, the formal role of the intelligence analyst supplements analysis conducted by officers for and by other citizens. Analysis at this level is successful even if an entire agency is not engaged since it depends most on individual initiative and situation-dependent measurements of success.

The implementation of problem-oriented policing is not as widespread as it could be. Herman Goldstein, the author of *Problem-Oriented Policing*, lamented its slow growth in his keynote speech at the International Association of Crime Analysts Conference in Seattle in 2005. The strategic thinking and analysis required to engage in true problem-oriented policing requires a shift in the core of traditional policing. Reflection on problems, rather than immediate action, suspending assumptions in order to examine facts from new perspectives, and the capacity to work on issues without obvious solutions over long periods of time — these qualities are needed to engage in problem-oriented policing.

The Police Foundation's "Problem Analysis in Policing" stresses the federal role in funding publication of problem analysis activities, including both what works and what does not:

If one is really committed to reforming police in this country, then acquiring and synthesizing the type of knowledge we are seeking and disseminating it to police departments are a high priority. Doing so would enable the police to act more intelligently with regard to specific problems. It would enable the police to be in a better position to describe to the public what they can and cannot do. That would equip both the police and the public to redefine the whole police operation in a way that makes what now often seems impossible possible. Being able to act honestly and intelligently, based on carefully developed knowledge, could help attract more people into the field that are challenged by the fact that this is a very complicated business; that policies and practices are based upon a broad body of knowledge and information that drives decisions.[110]

There are signs of new ways of addressing the changes needed to facilitate system-wide innovations in policing. The emergent position of "strategic manager" in law enforcement can facilitate such changes.

Strategic management is a systems management approach that uses active leaders in the organization to move change across organizational boundaries. A small team of personnel is assembled to analyze operational functions, identify inefficiencies, review systems integration, and detect gaps in management communications that hinder performance. In identifying organizational barriers, whether they are operational or caused by human dynamics, strategic managers are able to recommend strategies to the police executive to improve operations and quicken transitions, while working with managers to soften human resistance to change.

Although the police executive has the vision, the role of guiding the agency toward organizational renewal and change is the responsibility of all managers. Major transformation in an organization cannot rest with one individual but should be guided by teams under the direction of strategic managers. Police executives should scan their talent pools for command and support staff members who have the expertise, credibility, and competence to get the job done. Working with the chief executive and top managers, strategic managers assist in expediting change by educating, training, and marketing the reasons for change to management staff to make the vision a reality. [111]

---

[110] Herman Goldstein of The Police Foundation, "Problem Analysis in Policing," COPS-USDOJ, 37 at *http://www.cops.usdoj.gov/mime/open.pdf?Item=847.*

[111] Kim Charrier, Strategic Manager, Phoenix Police Department, Arizona, in "Strategic Management in Policing: The Role of the Strategic Manager," *The Police Chief* 71, no. 6 (June 2004) at *http://policechiefmagazine.org/magazine/index.cfm?fuseaction=display_arch&article_id= 324&issue_id=62004.*

Despite the imperfections and inadequacies in the processes mentioned in this chapter, analysts and experts value the focus on analysis implicit in each process, and there is general consensus that these processes are valuable and will evolve.

Next

The advent of the personal computer and software advancements have made the intelligence analyst's job more central to law enforcement. The following chapter explains how information and technology combine to promote an analyst's capabilities.

Chapter 6

WHAT WORKS: INFORMATION AND TECHNOLOGY[112]

Technology can't be the master — we should be the master. We utilize the technology to answer OUR questions. The technology should not replace what we do — it should supplement, enable, and speed our work.

It is a lot better to do good analysis on incomplete data than bad analysis on complete data.

The janitorial work of crime analysis is cleaning up the data.

The integrity of your personnel and their reporting is a core value to keep; we must have honest, sincere civil servants filling out reports to the best of their ability and duty.

Telephones were technology that changed our work — the capacity for crime reporting via 911 as well and a change in the types of crimes that occur, such as telemarketing fraud and facilitating communications between criminals.[113]

It is a capital mistake to theorize before one has data. Insensibly one begins to twist facts to suit theories, instead of theories to suit facts.

—Arthur Conan Doyle

Technology allows analysts to do in hours many things that previously took weeks. Armed with various software programs, the average analyst, with the right data and computer equipment, can map crimes, create an association chart, write a report and distribute it to every member of the force, create a wanted bulletin, and contact a colleague across the country, all in a matter of hours.

Computers and Information Systems

According to the U. S. Bureau of Justice statistics:

- Twenty-eight percent of local police departments in 2000, and 33% of sheriffs' offices, used computers for inter-agency information sharing. This includes three-quarters of all local departments serving 250,000 or more residents, and more than half of all sheriffs' offices serving 100,000 or more residents.

---

[112] *A Resource Guide to Law Enforcement, Corrections, and Forensic Technologies,* published by the Office of Justice Programs and Office of Community-Oriented Policing Services in May 2001, provides in-depth resource information. It is available online at *http://www.ncjrs.org/pdffiles1/nij/186822.pdf.*

[113] This and quotes above from interviews.

- In 2000, 75% of local police officers and 61% of sheriffs' officers worked for an agency that used in-field computers or terminals, compared to 30% and 28% in 1990.

- In 2000, 75% of local police departments and 80% of sheriffs' offices used paper reports as the primary means to transmit criminal incident field data to a central information system, down from 86% and 87% in 1997. During the same time period, use of computer and data devices for this purpose increased from 7% to 14% in local police departments and from 9% to 19% in sheriffs' offices.

- The percentage of local police departments using computers for Internet access increased from 24% in 1997 to 56% in 2000. Among sheriffs' offices, 31% used computers for Internet access in 1997, increasing to 67% in 2000.[114]

Even with the progress since 2000, it is likely that interdepartmental information sharing has not yet come to all local law enforcement agencies. Paper reports continue to be the main mode of data collection in the field. Without basic equipment and technology at the ground level of law enforcement, as well as the training to use the technology, how can we expect law enforcement data to be analyzed in an efficient and timely manner?

The Role of Planning in Technology

Effective planning for the integration of technology into policing, not only for crime mapping but for other analytical processes, requires insight into the information and *intelligence* needs of law enforcement. What is important to know? What new things can we see with all the data that was never possible to see before? Like designing a building, designing a technological infrastructure means that one must have a foundation to build upon, know the purpose of the building, and create an environment that supports the inhabitants' needs. Often technology is purchased and implemented without such foresight.

Implementing and integrating mapping technology with standard departmental procedures is not a simple or automatic process. Just purchasing the necessary software and computer equipment is not enough to ensure the implementation of a successful mapping strategy in a department; there needs to be a push to successfully implement a GIS plan. This suggests the importance of technical assistance and training in the development of successful mapping programs. Without such support and assistance, local departments are not likely to be able to successfully integrate computer mapping into problem solving and community policing.[115]

---

[114] *http://www.ojp.usdoj.gov/bjs/sandlle.htm.*

[115] The Police Foundation, Crime Mapping Laboratory, *Integrating Community Policing and Computer Mapping: Assessing Issues and Needs among COPS Office Grantees*, February 2000. See *http://www.policefoundation.org/pdf/CD9.pdf.*

Rather than having purchased technology forcing an agency to adapt to its structure, the technology should be designed to meet the needs of its users and be flexible to adapt as needs change and as new needs are identified.

Example of a Common Information and Technology Problem

Even though the NYPD is more advanced than most LE agencies in numerous ways, information and technology obstacles common to too many other U.S. police departments also linger at this key agency. The passage below describes the senseless situation of too little accessible information on outdated technological tools experienced not only in New York City, but also by many law enforcement entities in our country.

Dispatchers — each responsible for patrols in at least two precincts — often possess information that they're just too busy to share, information that could save a life; clues about what lies ahead, such as the name of the person who called for help, a description of an attacker, and whether guns or knives have been reported at the scene. Police officers in other cities get all this information, and more, from in-car laptops tied to sophisticated computerized dispatch systems. Police officers in New York, on the other hand, are thankful if their car has a terminal that can just spit out information about who owns a particular vehicle, and whether there are warrants outstanding on that person. [116]

Because the media now often portray law enforcement agencies as being armed with a multitude of high-tech tools providing them with instant information, policymakers are not pressured by the public to fund what is needed: the comprehensive overhaul of a technically outdated police world.

Issues in the Development of Technology

The reality is that the criminal world is outdistancing the law enforcement world in its adaptation to potentials of technology.

This trend has several aspects. (1) Law enforcement agencies are not able to afford to keep pace with the rapid change in technology, thus always leaving it behind the curve. Government agencies were built on the predicate that pieces of equipment (computers and their corresponding software are considered equipment) last as long as an employee. Computers are [inventoried like] desks, and until a realistic approach regarding keeping up-to-date with the change in technology is taken, law enforcement and its intelligence component will be inferior (technologically) to the criminal organization. (2) Although the explosion in technology generally allows the intelligence units to do their duties more efficiently and faster, there is a danger in believing it replaces analytical thought, investigative intuition, or genuine hard work.[117] (3) This technology explosion will allow criminal organizations, since they

---

[116]Deborah Gage and John McCormick, "The Disconnected Cop," *Baseline Magazine* (online), 10 September 2002, *http://www.baselinemag.com/print_article2/0,2533,a=30779,00.asp.*

are unrestricted by bureaucracy and funding, to outdistance law enforcement's ability to investigate or unearth specifics about their criminal acts. (4) Computers, the Internet, and public information sources all contribute to making the unsophisticated criminal more sophisticated.[118]

The Data Themselves

Data are the basis for all analysis.

Data integrity is universally critical, because *every part of the analytical process that follows depends on the data on which it is based.*[119]

Criminal and intelligence analysis are more likely to be effective when the information (data) is *available* and when it is of the *highest quality*. Not only must data be accurate and reliable, they also must be properly formatted. The well-worn saying "garbage-in, garbage-out" does apply to law enforcement analysis. Many analysts interviewed expressed a desire for better information in terms of improved collection, improved data entry, and improved accessibility of data. Analysts who have information systems that worked well were certainly the happiest in their role as analysts.

Crime and disorder data contain many errors. Much crime is not reported, so the biggest data error is not having known about events that occurred. Other events are not accurately described. One might not know about important characteristics of the event — what it is, when it occurred, or where it happened. These errors can confound analysis and preclude effective action. And the confusion arising from errors in crime and disorder data becomes compounded when we compare across jurisdictional borders. As important as data quality is for normal crime analysis work, it is even more critical when examining data from different jurisdictions that may suffer from different rates of error. Consequently, they require serious attention.[120]

Collection of information by police officers for analysis is in a dismal state, according to a number of participants. Local-level policing, unlike investigations at state and federal levels (where analysts may work several years on a case), is a high-volume business with a high volume of data. Huge amounts of data are collected every day in large agencies.

---

[117] If a person leaves his or her job and takes the computer home, the agency will file charges for stealing the computer, all the while not realizing that the most valuable information is located in the person's head.

[118] Martin A. Ryan, "The Future Role of a State Intelligence Program," in *IALEIA Journal 12*, No. 1 (February 1999), 28.

[119] IACA, *Exploring Crime Analysis*, 144.

[120] John Eck, "Crossing the Borders of Crime: Factors Influencing the Utility and Practicality of Interjurisdictional Crime Mapping," *Overcoming the Barriers: Crime Mapping in the 21st Century* (National Institute of Justice, Police Foundation, 2002), 8-9. See URL *http://www.policefoundation.org/pdf/barriers-eck-2002.pdf.*

Beyond collection problems, once the information is entered into computerized systems, data problems often arise.

Examples of Data Problems[121]

Empty Fields	Aliases
Typographical Errors	Malapropisms
Punctuation	Generalizations
Abbreviation	Invalid Entries
Omissions	Extraneous Errors

These errors occur in many automated information systems and interfere in the analysis process. Searching data with such mistakes is an analyst's nightmare — it makes the work so much more difficult because errors must be corrected before analysis can be done.

Temple University professor, former Australian detective, and participant in the present study Jerry Ratcliffe, presents examples of these and other related problems and their ramifications in a forthcoming journal article.[122] The lack of trained data entry personnel affects the quality and consistency of data. Lack of audit procedures and capacities contribute to data disasters; in Ratcliffe's study, an officer who had done an internal audit found *a fifty percent error rate in data recording*.[123] Selective data entry can also be a problem. Certain crime types may be recorded in greater detail, a factor that aids analytical efforts, while others may capture minimal information.[124]

Obviously, if we do not have audit practices in law enforcement agencies, we have little idea of whether the information is correct, complete, and entered in a fashion that facilitates analysis. We often make inferences based on data that are incorrect, without the ability to ascertain their accuracy. Aggregate crime statistics reported on the national level are affected by this problem. Auditing and audit standards are not universal.

Without quality crime data, we cannot measure changes in the criminal environment.[125] Nor can we measure the effectiveness of police work. Thus, several analysts interviewed for this study expressed a desire for better data.

---

[121] *Exploring Crime Analysis*, 146.

[122] Jerry H. Ratcliffe, "The Effectiveness of Police Intelligence Management: A New Zealand Case Study," forthcoming in *Police Practice and Research*, and available at *http://www.jratcliffe.net/ conf/Ratcliffe%20(in%20press)%20NZ%20case%20study.pdf.*

[123] "The Effectiveness of Police Intelligence Management."

[124] "The Effectiveness of Police Intelligence Management."

[125] "The Effectiveness of Police Intelligence Management."

Collection Priorities and Analysis

Mark Lowenthal, an authoritative observer of intelligence practices observes that:

analysts should have a key role in helping determine collection priorities. Although the United States has instituted a series of offices and programs to improve the relationship between analysts and the collection systems on which they are dependent, the connection between the two has never been particularly strong or responsive.[126]

Similarly, analysts in local-level law enforcement are faced with a variety of discontinuities between collected data and their analysis of it. Traditionally, the data collected by officers was filed in drawers and not easily searchable. Officers with good memories and an interest in linking crimes could uncover patterns. Some series and patterns are obvious without resort to examining computerized data. However, officers work different shifts and it is next to impossible to have a global view of the incoming information in a large agency unless data are digitized.

Unless police reports have drop-down boxes to force standardization of electronic data entry, officers enter qualitative variables in their own language. Was "red" actually: burgundy, rust, or maroon? Data in reports also rely on the reporting accuracy of the person filing the complaint and the time allotted to the officer to write up the details, as he may get another urgent call for his attention. Some reports have detail; some do not, due to the fluid nature of police work and gray standards in data collection.

This is a real problem affecting the ability to spot and assess a potential crime series. There may be a serial rapist who rapes only once a month in various jurisdictions. Let's say that he always approaches victims in an elevator. Even if these data are shared across jurisdictions, if the MO does not mention the elevator as significant, it may not be possible to see a serial rape pattern unless there is physical evidence to link the crimes.

Example of Success: Automation to Streamline Work

Automating traditional analytical tasks is effective in freeing analysts to do more relevant work than making standard statistical charts and graphs. In Washington State, Pierce County Sheriff's Department Crime Analyst John Gottschalk and other analysts on his team had the technical skills needed to create an automated monthly COMPSTAT-type Microsoft PowerPoint presentation (example on the next page).

---

[126]Mark M. Lowenthal, *Intelligence: From Secrets to Policy* (Washington, DC: CQ Press, 2003), 47.

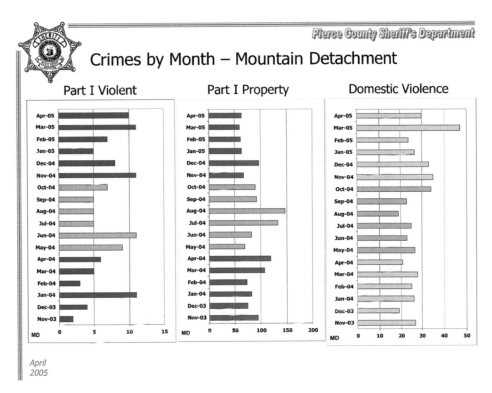

Pierce County Sheriff's Department

Crimes by Month – Mountain Detachment

| Part I Violent | Part I Property | Domestic Violence |

April 2005

Source: Courtesy of John Gottschalk.

Analysts need technical support to automate their work if they themselves do not have such skills. An analytical team whose members have different strengths can exploit those differences to find the most effective methods to streamline work. Automation is technically possible in a number of analytical tasks that formerly required days to accomplish.

Automation of routine work is not the cause of success; rather, it frees analysts to do more exploring of data. Most crime analysts prefer to read crime reports and automation helps them find those reports in a records system much more easily than in hard copy, especially as they can be queried by a number of variables, such as crime type or locations. Another type of automation is the use of templates. The Police Foundations Crime Mapping Laboratory offers "Crime Analysis and Mapping Product Templates" in an electronic version.[127] Although such templates have limited utility in terms of originality and creativity, they serve as a guide for standardizing products within an agency and within the profession of law enforcement analysis.

---

[127] Available at *http://www.policefoundation.org/docs/mapping_templates.html.*

Examples of Success: Analysis Information to Officers in Their Cars

Address http://intranet.alexgov.net/apd/Crime/capage.htm

Links CNN.com | Google | IACA | NIJ's M.A.P.S. | Virginia Crime Analyst Network | Virginia State Police | washingtonpost.com - News Front | CMAP

| APD Most Wanted | Bulletins | Statistics | Maps | FAQs | Sex Offender Information |

Peeping Tom/Burglary
Cases in the Hamlets
04/08/05

Larcenies from Auto at Alexandria Apts
UPDATED 04/20/05

Larcenies from Autos on Stevenson Ave
04/08/05

Spoiler Thefts - West End
03/29/08

Thefts from Laundry Systems
03/17/05

Taxi Cab Crimes
(UPDATED 03-29-05)

CLICK HERE FOR REGIONAL CRIME BULLETINS

Who We Are and Where to Find Us:

Note: Contact information removed by editor.

Crime Analysis Unit Web Page Available in Police cars in Alexandria, VA.

Source: Courtesy of Mary Garrand, analyst with the Alexandria Department.

Bringing information down to the workspace of officers on the street is another form of technology that works. Intranet web pages that allow officers to access analytical information from a crime analysis unit, mapping applications that allow officers to analyze data in their cars, and the ability to send bulletins to officers on the street directly from the analyst to the officers are examples of effective use of technology.

WHAT WORKS

Field Reporting

The ability to transmit data directly from the field to automated systems represents an effective advancement in the field of data collection. Agencies that have officers reporting from police car to computer systems help analysts obtain timely information and increase the likelihood of their producing relevant analytical products. For example, officers who have field contact with a suspicious person, and who enter the contact information into a system, make it available for the analyst who is trying to find potential suspects for investigators of a crime series.

Data Mining

We are getting better at exploring data through formulating questions but we are just touching the tip of the iceberg. The concept of data mining is growing — and not only automated data mining.[128] The analyst can become skilled at developing and implementing ad hoc queries. Knowing what kinds of questions to ask of the data tends to maximize their utility and demonstrates the skills of an experienced analyst.

Ad hoc queries based on the analyst's familiarity with common problems specific to the jurisdiction increase the efficacy of data mining. For example, certain types of commodities may be particular to an area, such as antique house fixtures which are stolen in parts of the U.S. Northeast for resale to home builders in other parts of the country. Rapes by unknown assailants are more likely to be serial in nature than those with known assailants, so a query based on the relationship of the victim to the offender can be valuable. A certain type of gun may be queried for its relationship with a known cache of stolen guns in order to see patterns of the weapons' distribution. Information about elderly victims might be queried if there has been a pattern of victimization related to traveling criminals who prey on the elderly. Ad hoc queries on a particular phrase used by a suspect in a crime, if captured in the crime report, may lead to identification of a series of crimes. The wide possibilities for ad hoc querying stand to help investigations by developing leads and exposing relationships in data that help achieve a more comprehensive understanding of a crime problem.

Technology an Analyst "Can't Live Without"

Mary Garrand, Crime Analyst Supervisor for the Alexandra PD, reports that crime analysts cannot live without the following technology:

- Database software to track crimes and offenders
- GIS (Geographic Information System) software
- SPSS (Statistical Package for the Social Sciences) or other statistical software
- Excel
- Word or other publishing software
- The Internet

Ms. Garrand notes that existing technology on many desktops often has a number of application extensions that users do not know about or do not know how to use. She adds that the software in the list above provides all the basic tools to do the work of a crime analyst.

Lois Higgins of the Florida Department of Law Enforcement provides the following software to attendees of the FDLE Analysts Academy:

- RFFLow charting software (an inexpensive flow-charting software)

---

[128] Interview subject Paul Wormeli of the IJIS (Integrated Justice Information Systems) Institute recommends that every analyst learn OLAP (Online Analytic Processing) to facilitate data mining.

- MapPoint mapping software (an inexpensive but limited mapping software — primarily valuable for pin mapping)

Law enforcement analysts also should have *training* in the use of mapping software, spreadsheet software, desktop publishing software, statistical software, database software, flow-charting software, and the Internet.

Access to Information

Often, access to information is assumed. Just because information exists does not mean it is accessible to the people who need it. In one story, officers had to go to the newspaper reporter they knew for access to commercial databases to find out more information on suspects they were investigating because their agency did not pay for such services. Public sector workers often do not have the same access to data resources as private sector workers. Subscriptions to these services remain costly, and often those in law enforcement who would benefit from them do not know the resources exist or cannot adequately articulate their benefit to superiors when they have the opportunity to request tools. If their agency has access to commercial databases, such as Lexis/Nexis or Choice Point, it is often limited — only one person in an agency may be able to use it.

Analysts in many law enforcement agencies have trouble getting access to information within their own departments, not to mention from outside of their agencies. Investigative files that could improve the depth and quality of analyses are kept locked away. Data on parolees, probationers, and even the roll of individuals arrested in their own jurisdictions may be a challenge to obtain. Data on truant students, juveniles, court dispositions of cases — and many other such data — may not be obtainable.

Analysts would like more and easier access to other types of government data. Land use, property owner information, and licensing data are just a few examples of government data that could be used to help understand crime problems in a more comprehensive fashion if correlated to crime data. If the local government owns the data, analysts ask, why not share it in a purposeful manner to help provide better services to citizens?

Correlating various data sets can be used in ways limited only by the imagination of the analyst. Sometimes the value can only be discovered through experimentation. For example, ortho-photographic overlays on pin maps may reveal that a serial pattern of arsons is occurring primarily at the end of streets next to boarded-up buildings. (An ortho photograph has consistent scale throughout and can be used as an accurate overlay of a map.) Such details may only be seen through ortho photography or by site visits to each crime location. Similarly, a detailed analysis of property ownership may indicate that vacant properties that are used as drug houses have the same owner.

Intimacy with Data

Knowledge of the data past and present in one's agency is a key to success. Having a deep understanding of the data and its evolution, a kind of corporate knowledge, is very useful. This type of familiarity is necessary for an individual to note anomalies and see analogies. For example, a new analyst may study a month's worth of robberies, create a report indicating that there were ten truck drivers robbed, and not know that this is abnormal for the jurisdiction unless he or she is familiar with the "norm." The same analyst in that month may note one robbery of a fast food restaurant and not see that it is exactly like a robbery that occurred a year ago, because he or she has no historical frame of reference. Familiarity with norms and having knowledge of past instances helps one recognize the unusual as well as the similar.

One expert said he likes to "smell the data — get on the ground level and see where it is coming from."

Next

The next chapter explores the concept of "analysis."

Chapter 7

WHAT WORKS: ANALYSIS

Investigating one individual responsible for 15 robberies is cheaper than 15 separate robbery investigations.

You have to be that nine or ten year old and ask a lot of questions.

Automation will make us obsolete unless we do analysis. [129]

Most importantly, plan to spend the bulk of your time on the analysis phase, which will give a much more detailed picture of the problem. Many times, a thorough analysis will yield surprising findings about the underlying causes of the problem. Often, preconceived ideas about appropriate responses will thus be changed or even abandoned for better ones.

— *Chula Vista Police Department,*
PERF Response/Assessment Survey, 1999:12 [130]

Strangely, each time we ask prevention practitioners, field researchers, or our students what analysis actually means we find a curious mix of answers. Some — the more academic — believe analysis is a form of research on some theoretical question regarding the causes of, or response to, crime. Others — the more practical — feel analysis is not a theoretical exercise in research but rather that it must solve a specific problem. They echo a twisted version of a Star Trek refrain: "Dammit Jim, I'm a problem solver not a number cruncher!" Why then, we wondered, do so many think these two views are mutually exclusive? [131]

The individuals interviewed for this project stress that analysis that works is imperative, but there was not a great deal of elaboration about what good analysis is nor how to do it.

"Analyst" as Title and as Word: "Analysis" as a Limitation

Analysts and experts interviewed for this study did not stress the technique of analysis in its classic definition — breaking things apart to study them — as the central activity of their work. Technically, analysis is only a minor aspect of their work. Analysts supporting a major investigation are actually doing the opposite — organizing and synthesizing snippets of information to form a larger picture. Analysts looking for serial crimes are identifying patterns in large data sets — linking information through analogy and spotting anomalies.

---

[129] Quotes from interviews.

[130] Debra Cohen, *Problem-Solving Partnerships: Including the Community for a Change* (USDOJ:COPS, June 2001), 4.

[131] Greg Saville and Chuck Genre, draft manuscript *Understanding Neighborhood Problems*, University of New Haven.

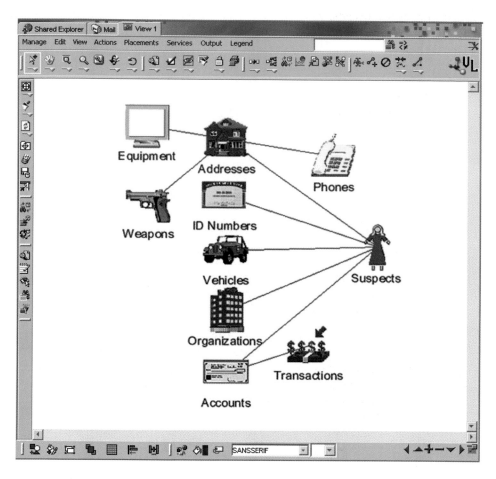

The Synthesis Behind Analysis.

Source: Chart courtesy of David Jimenez, U.S. Border Patrol analyst

Gary Klein studied naturalistic decisionmaking by chronicling the strengths useful in making difficult decisions.[132] His chapter on "The Power to See the Invisible" was reflected in the accounts of several analysts who value their ability to "see what others do not see." This is an important strength, and Klein's insights on this ability may be useful in understanding the characteristics of this possibly crucial analytical capability. Among

---

[132] Gary Klein, *Sources of Power: How People Make Decisions* (Cambridge, MA: The MIT Press, 2001), 1.

the many things that experts, with their tacit knowledge, can see, but that are invisible to everyone else:

- Patterns that novices do not notice.
- Anomalies — events that did not happen and other violations of expectancies.
- The big picture (situational awareness).
- The "way things work."
- Opportunities and improvisations.
- Events that either already happened (the past) or are going to happen (the future).
- Differences that are too small for novices to detect.
- Their own limitations.[133]

The Human Factor, Technology, and Artificial Intelligence

Although technology offers a gateway to analyzing enormous amounts of data in ways previously unimaginable, many analysts emphasized the importance of human beings who read reports and connect different sorts of data to uncover patterns and solve problems. The following passage explains why this observation remains valid.

[On] the question of whether expert system, neural net, or artificial intelligence approaches would provide a better means of inferring patterns from such spatial data. We doubt it; these approaches may be very useful when trying to look for patterns of, for example, a disease, where the essential aspects of the disease manifest themselves in different cases in similar ways. This is not the case in crime analysis, unfortunately, because each burglar may have a radically different modus operandi (MO). Training a computer algorithm to recognize burglary patterns from the MOs of burglars 1 through 99 will probably not be of much benefit in having it recognize burglar 100. In other words, to a computer analyst a "pattern" is a condition to be diagnosed (like jaundice) while to a burglary detective a "pattern" is not so much a condition (like burglary) but an MO that relates to a specific offender. [134]

Understanding aggregate crime problems, which may be uncovered by automated data mining, leads to the development of new strategies to prevent specific types of crime, but in order to identify a specific individual who is committing serial crimes, or a particular crime, one must factor in many variables, including changing contexts. The robber who only strikes in dry weather because he does not have a car, who says different words to his victims but always in a polite tone of voice, who wears a bandana or a scarf or a muffler,

---

[133] Klein, 148-149.

[134] Marc Buslik, Chicago Police Department, and Michael D. Maltz, University of Illinois at Chicago, *Power to the People: Mapping and Information Sharing in the Chicago Police Department*, n.d.,129, *http://tigger.uic.edu/~modem/Pwr2Ppl.PDF*.

depending on one's interpretation — these nuances require human perception to discern possible significance.

Analysis Tasks: A Description from a Centre for Analysis

The Police Service of Northern Ireland offers a description of key analytical tasks found useful for law enforcement analysts:

> Traditionally key tasks for crime analysts have included examining information on "hotspots" and identifying patterns that might indicate crimes are linked. Some crime is seasonal or occurs at particular times of the day or week. By examining information properly it is often possible to provide advice on how, and where, to best deploy operational personnel.

> "Live time" plotting of hotspot trends in the recovery of stolen vehicles, for example, allows police to immediately target areas where problems are most likely to occur. Intelligence analysts help in tackling prolific criminals by drawing together lots of information to identify networks and opportunities for police action. Police analysts can also provide the detailed information necessary to support multi-agency crime prevention/reduction initiatives — for example a detailed analysis of all alcohol related assaults in a particular area can provide the essential catalyst to concerted action by hospitals, health/education boards, licensed premises and others working with the police.

> Skilled analysts also play a role in assisting major crime investigations. Often they are used to sift through large amounts of information. Charts and accompanying reports — which depict visually "what happened" — are routinely produced in murder and other serious cases. They can be very useful, particularly in complex investigations, where statements need to be checked and corroborated. And they can assist busy detectives to identify new lines of enquiry. In court they can provide invaluable help to prosecution and defence council and ultimately to the judge and members of the jury. [135]

TYPES OF ANALYSIS

Law enforcement analysts are dependent on the vision of their agency, their access to information, skills and training received, the objectives set for them and, in instances where managers do not understand their role, on objectives set by the analysts themselves.

Law enforcement analysts engage in some or all of the following analytical tasks, depending on the mission of their organization, their assignment, the availability of resources, and their skill levels and the specific training they have received.

---

[135] See http://www.psni.police.uk/index/departments/analysis_centre/pg_why_analyse.htm.

■ Frequency Analysis — the quantity of disorderly behavior and criminal variables

Frequency analysis is the most common task of crime analysis. Counting crime, comparing counts, and analyzing crime trends are examples of frequency analysis. Besides crimes, 911 calls for service may be counted, and frequencies of other variables such as types of vehicles, color of suspects' clothing, and the like may be recorded in order to find patterns in data. Some agencies have software that allows users to set thresholds of crime or calls so that, if the frequency exceeds the threshold, the analyst or officer is alerted to direct attention to the specified area.

■ Spatial Analysis — the "where" of crime/disorder

Spatial analysis is another common task of crime analysis. Crime mapping, whether using pin maps or sophisticated Geographic Information System (GIS) applications, aids in crime analysis to identify patterns of location. Spatial analysis can include refinement to specific locations in buildings, such as hallways, certain rooms, elevators, and parks if data are collected to that level. Spatial analysis also includes examining the spatial relationships between variables, such as ATMs to robbery, vandalism to schools, etc. Innovators in spatial analysis have adopted ideas from other disciplines, such as animal movement, to study the predatory behavior patterns that sometimes seem to apply to certain types of criminal behavior.

■ Temporal Analysis — the "when" of crime/disorder

Some crimes occur in distinct patterns of time. The more difficult challenge is to see patterns that occur over years even in the same jurisdiction. Humans, even analysts, tend to think in short-term time frames. A rapist who strikes every other year is less likely to be detected as a serial offender through traditional temporal analysis, compared to one who strikes every month. Temporal analysis identifies patterns of crime/disorder by time of day, day of week, time of month, time of year, season, special events (holidays, sporting events) as well as the tempo of crime.

■ Modus Operandi Analysis — the "how" of crime/disorder

Modus operandi (M.O.) analyses is especially helpful in highlighting whether crimes are related. Capturing MO data in standardized, digital format is a most difficult challenge. Modus operandi analysis consists of analyzing the qualitative aspects of a crime or disorder incident or collection of incidents by examining the method used in committing the offense. Data include approach method, phrases spoken, entry method, degree of force, specific types of items taken, weapons used, and types of graffiti, which are then analyzed to uncover relationships between specific crime and/or disorder events. Qualitative data can be coded to aid the efficient exploitation of data. Within this analysis, in some instances, the "why" of crime may also be inferred.

■ Suspect/Victim Analysis — the "who" of crime/disorder

Suspect description information is analyzed to identify crime series (crimes perpetrated by same individual(s)) or trends (such as emerging gangs). Victim information is analyzed to uncover series (such as those targeting the elderly) and trends (such as an increase in juvenile victims).

- Property Analysis — the "what" of crime/disorder

Property taken in a crime can be analyzed to uncover trends or to link related crimes. For example, pawn shop data can be analyzed to uncover criminal activity; types of stolen vehicles can be studied to find which cars need theft prevention modifications.

- Problem Analysis — the depth of crime/disorder

Problem analysis constitutes a multi-faceted approach to analyzing a crime or disorder problem, and may require site visits and interviews with members of the community. Crime analysts may also contribute to the study of individual problems such as automobile traffic, automobile accidents or domestic violence.

- Demographic Analysis — the population and crime/disorder

Demographic analysis focuses on demographic variables and their possible influence on crime and disorder in the community.

- Market Analysis — the commerce of crime

Market analysis addresses recovered guns, stolen goods, and illegal narcotics activity, for the purpose of uncovering markets and developing strategies to disrupt the markets.

- Network/Association Analysis — the organization of crime

Network analysis discovers and displays the relationships in a criminal group and the activities of specific criminals.

- Communication Analysis — uncovering the relationships among suspects

Communication Analysis explores the relationships of suspect individuals/organizations as exhibited through telephone/computer records.

- Financial Analysis — transactions of suspect individuals/organizations

Financial analysis scrutinizes the records of suspect individuals and businesses (or other entities) to uncover relationships and possible illegal activity.

APPLYING ANALYTICAL APPROACHES

Describing law enforcement analytical techniques in any depth is beyond the scope of this book. Some of the resources cited in the book can assist readers who wish to explore this topic in greater depth. To help in understanding how analysis is applied, general examples of the application of analytical approaches follow, using the analysts' story types from Chapter Two and the processes from Chapter Five as a framework.

- ***The Identification of a Crime Series Story:*** A crime analyst is the first one to notice an existing pattern of crimes in which the same perpetrator(s) seems to be responsible — a crime series — and the series does, in fact, exist.

In this type of success, an analyst will note that more than one crime (frequency) has similarities in MO. MO is the most important variable in determining if a crime is part of a series. *How* a crime is committed is the indicator of individual behavior, and because a series involves the same individual or group of individuals, MO analysis is most important in this scenario. Serial crimes often have a spatial relationship, and may be found through analysis of crime hot spots, but they also may occur across large areas and may not be discovered through spatial analysis. The time (hour of occurrence) of a crime plays an important role in predictive analysis. Some (but not even most) serial criminals commit crimes in very distinct patterns of time, increasing the likelihood of accurate predictions. Analysis of suspect information, including a physical description, often is the key to discerning a series, but witness descriptions of offenders are often poor. Victimology is also a key in linking some serial crimes — this means analyzing similarities in victims. This approach may also include the question of what type of business is targeted in property crime.

- **The Pieces of Information Turning Into a Big Case Story:** An intelligence analyst gets boxes of information, sometimes CDs full of information, and sorts through all of it (analyzes it) to uncover information that leads to an even bigger investigation.

In this type of success, the analyst compiles a picture of a problem using information on associates of a suspect (which may be a person, a business, and/or a group of people), possibly through communications such as correspondence, telephone records, email, as well as through surveillance and informants' reports. Financial analysis to find possibilities of money laundering or other illegal income indicators is conducted through analysis of bank records, charge receipts, assets and liabilities information, and other relevant evidence. Market analysis of the flow of commodities (such as guns, drugs, legitimate or counterfeit goods) may indicate relationships involving illegal activity. The analyst pulls clues from available information to find the important leads and help direct the activities of investigators in their search to find more puzzle pieces for the case.

- **The Prediction Leading to Arrest Story:** A crime analyst makes a prediction regarding the likely time and place a serial criminal will next offend and the offender is apprehended based on the analyst's accurate prediction.

This type of analytical work is highly dependent on frequency, temporal and spatial analysis. Although, for statistical accuracy, a number of crimes must have occurred to make a prediction that meets scientific standards, analysts make predictions on fewer crimes in their effort to prevent another crime. They do not wait for a crime to occur if there is enough information to make an informed forecast. Analysts use temporal analysis to predict the most likely time the next crime will occur. There is no standard formula to do this — analysts are like artisans in their forecasting techniques — they use a variety of approaches. The same principle applies to spatial analysis. The location on a map is part of the forecasting, and this is enhanced if the analyst is very familiar with the territory and potential targets. Using environmental survey techniques by going into the field to assess the similarities of targets, or of places crimes have occurred, helps the analyst decide the potential of a future target being hit.

- *The Successful Investigation Leading to Prosecution Story:* An intelligence analyst supports an investigation through appropriate analysis and visualization of data, creating relevant reports and graphics; the analyst's work is used in court to help successfully prosecute the targets of the investigation.

This type of success often involves network analysis, event flow analysis and timelines. It involves ongoing organization of information for investigators. It may involve maps to visualize criminal activity and/or the flow of commodities. The analyst works as part of a team and helps during the investigation and through prosecution with documents appropriate for the courts.

- *Intelligence-Led Policing*

The Intelligence-Led Policing model employs all types of analysis at every level of policing. Implementation of this model has occurred chiefly in the United Kingdom and has been proposed for the United States, but it is not yet applied in a systematic way in this country.

- *COMPSTAT*

COMPSTAT employs the technical products used by analysts, such as frequency charts, crime mapping, and other types of statistical and graphical data, but by itself, it is not analysis. It is primarily used as a management accountability tool. Too often, COMPSTAT is confused with crime analysis. The NYPD has a separate crime analysis section as well as a COMPSTAT process in place.

- *Community Policing*

Community policing employs a number of analytical approaches. If there is a crime trend identified through frequency analysis, citizens may be warned in order for them to engage in crime prevention measures. In some jurisdictions, crime maps and statistical data are available on the Internet so that community groups and individual citizens may be fully informed of local crime problems. Addresses are usually rounded to the hundred-block number to protect citizen privacy in the tabular data. Community police officers may work with citizens on chronic crime problems using analytical support to identify the extent of the problem. Demographic analysis, analysis of land use and property owner information, crime and 911 call data, environmental surveys, and interviews of stakeholders may be part of community policing analytical support.

- *Problem-Oriented Policing*

Problem-oriented policing analysis depends on the nature of the problem selected for study. Background analysis of area demographics, crime hot spots, the baseline of past and current policing strategies, interviews with stakeholders, as well as frequency, spatial, and temporal analysis are all important in this approach to analysis. It generally requires a higher level of research skills and long-term, dedicated resources to carry out adequately and effectively.

VIGNETTES OF SUCCESSFUL CRIME ANALYSIS

The following examples of analysis "success" illustrate the application of a variety of analytical techniques specific to crime analysis.

1. (From analyst Anne Gunther, Hampton, VA) The analyst took a Crime Line call regarding the identity of a subject who had committed a "robbery" of a commercial establishment. It was known that the robbery was actually a burglary (citizens often use the incorrect terminology). There had been several with the same MO. The name was queried and it was learned that the subject had a girlfriend who lived in the vicinity of the burglary locations, and he had previously committed the same type of act. The subject rode a bicycle and the items stolen were easy-to-carry entertainment items. After the bulletin was disseminated, the subject was field interviewed. He was arrested.[136]

2. (From analyst Doug Rain, Aurora, CO) Over a period of a few weeks, I identified what appeared to be an emerging crime pattern involving car break-ins in apartment complexes. The pattern was occurring within a four-hour period on only Sunday and Monday nights at apartment complex parking lots which backed up to greenbelt areas. I wrote a pattern report, which included an action plan, and gave it to the patrol commander. The commander took action by assigning a team of two uniformed officers to "stake out" the parking lot, which I had projected to be the next target. On the first night out, the team observed a young man climb over the privacy fence separating the apartment parking lot from the adjacent greenbelt. The subject looked around, approached a parked vehicle and pulled a screwdriver from his pocket. He then pried the driver side wing vent and stuck his arm through the opening in order to unlock the door. While his arm was still inside the victim's vehicle, the officers came up behind him and arrested him. He nearly fainted with surprise. After the arrest, the break-in pattern ended.... We were having a rash of robberies at dry cleaner establishments. The suspect description was the same in each one — he wore a Halloween mask. Additionally, the suspect seemed to be getting bolder with each robbery. No solid pattern existed for day of week or geographic location but all the robberies occurred within a one-hour period of the day. I did a projection (forecast) as to where and when the next likely robberies would occur. The projection was based on the suspect's tendencies rather than a tight pattern. The SWAT team and robbery unit supervisors were given the projection report. They took action by conducting surveillance on the two most likely dry cleaner locations identified in the report. The first day produced no results. However, on the second day the SWAT team observed an individual matching the general physical description of the suspect approach the cleaners at the projected time. Just as he reached for the door to enter, he pulled a Halloween mask from under his coat, put it on and entered the business. From their surveillance point, the SWAT team had the business covered in a matter of seconds. When the suspect exited the

---

[136] Deborah Osborne and Susan Wernicke, *Introduction to Crime Analysis: Basic Resources for Criminal Justice Practice* (Bingingham, NY: The Haworth Press, 2003), 108-109.

cleaners, he was arrested immediately outside on the sidewalk without incident. With his arrest, the pattern ceased.[137]

3. (From Lt. Tom Evans, Pinellas County, FL) Our Robbery — Crimes Against Property — Unit used analysis reports in several significant cases. One case involved a series of regional armed robberies involving restaurants. In this case, the analysis aided in obtaining complete confessions from the robbery suspects who previously denied involvement in the crimes. In this case the detective invited the analyst to sit in the debriefing interview after the suspect confessed. The analyst was able to refresh the suspect's memory on a number of cases. The second case involved armed robberies of local pizza delivery shops. In this instance, the analyst produced a forecast of the day, date, time, and location of the next robbery. I am pleased to say the detectives' surveillance of the business, based on the forecast, resulted in the suspect's arrest. In a third case, an analysis of a residential burglary series suggested that the perpetrator was using the public bus system for transportation. Based on the analyst's suggestion, detectives conducted a surveillance of bus stops in the forecast target area. Although another lead led to the suspect's arrest, during his confession to the crimes, he admitted to using the county bus system for transportation and on his last bus trips to commit a burglary he spotted detectives' surveillance and got back on the bus. We have many more like these...[138]

4. (From Brian Cummings, Richmond, VA) One Richmond Crime Analyst identified a pattern of commercial robberies in the downtown area of the city of Richmond in 1996. The robberies started with one in January 1996, followed by one in February and two in March and none in April. Between May 5th and May 7th, there was one each day. The robberies continued to occur with greater frequency through the month of May. The analyst had issued several reports during this time. The analyst identified physical descriptions along with various MOs. The analyst also included information obtained from a sergeant at Third Precinct concerning a possible suspect with a similar physical description. One of the MOs that was identified was that one location was targeted only on the weekends (Friday-Sunday) on the evening/midnight shift. This location was near a major hotel and conference center. The analyst noted that the robberies were in an area that would allow the suspect an easy get-away route and that some of the locations had been targeted more than once. It was also noted that the suspect would leave the scene in a northern direction. Investigation subsequent to the capture of the suspect revealed that the suspect lived to the northwest of the Downtown area. The detectives used the information provided by the analyst to set up surveillance and an apprehension was made. The subject was arrested and was the same individual about whom the Sergeant from Third Precinct had provided information during the investigation.[139]

---

[137] Osborne and Wernicke, 109.
[138] Osborne and Wernicke, 111.
[139] Osborne and Werncike, 111.

5. (From Crime Analysis Manager Gerald Tallman, Overland Park, KS) Overland Park was hit with a marked increase of residential burglaries through open garage doors. These "Garage Shopping" incidents were primarily centered in the southern half of our city. In a unified effort of Crime Analysis, Crime Mapping, Patrol, Investigations, and COPPS (Community-Oriented Policing and Problem Solving), we were able to successfully educate our citizens to keep their garage doors closed and locked, and within a short time frame, were able to almost eliminate these incidents. In a speech on crime mapping and technology, U.S. Attorney General Janet Reno publicly praised our use of technology and cooperative policing efforts in combating this problem.[140]

The law enforcement analyst is generally expected to analyze data and information in a variety of ways to support the following law enforcement activities:

- Management Decision Making
- Patrol Deployment
- Community Policing
- Problem-Oriented Policing/Problem-Solving Policing
- Intelligence-Led Policing
- Allocation of Resources/Request for Resources
- Investigations
- Apprehension/Arrest
- Prosecution
- Grants
- Crime Prevention
- Problem Solving
- Evaluation

WHAT WORKS

Crime Mapping

Federal funding to promote crime mapping in local law enforcement has resulted in a broad spectrum of agencies having crime mapping capabilities.[141] Sometimes agencies confuse having a crime mapping capability with analyzing information. Simply producing maps becomes the analyst's job, rather than interpreting the meaning of the information on the maps.

---

[140] Osborne and Wernicke, 111-112.

[141] This type of federal funding may have exaggerated the role of mapping in analysis — agencies acquired the ability to map crime but not the knowledge of other types of analytical tools, skills, and strategies. Funding to develop the profession (or vocation) of analysis itself, within law enforcement agencies, is not available.

It is easy to underestimate or overestimate the rate of change and the long-term impact of technological changes in policing — including crime mapping. While current and future advances promise to lend substantial support to law enforcement, we should remember that technologies such as crime mapping are only tools and, like other tools, their benefits to society depend on the humans who wield them.[142]

The author's interview subjects described analysis using Geographic Information Systems as effective when used in conjunction with other types of analysis, including temporal and MO analysis. Many online resources exist to provide a broader understanding of crime mapping issues in law enforcement.

Online Resources for Crime Mapping

These online resources will allow readers to investigate crime mapping more thoroughly:

Mapping and Analysis for Public Safety (MAPS)
http://www.ojp.usdoj.gov/nij/maps/

The Police Foundation: Crime Mapping News
http://www.policefoundation.org/docs/library.html#news

Advanced Crime Mapping Topics
http://www.nlectc.org/cmap/cmap_adv_topics_symposium.pdf

Mapping Crime: Principles and Practice by Keith Harries
http://www.ncjrs.org/html/nij/mapping/pdf.html

Law Enforcement Agencies with Crime Mapping
http://www.ojp.usdoj.gov/nij/maps/related.html

Overcoming the Barriers: Crime Mapping in the 21st Century — an occasional series on the human and technological barriers that police agencies face in the implementation and integration of crime mapping
http://www.policefoundation.org/docs/library.html#barriers

Source: Author.

Geographic Profiling

Geographic profiling is a tool used by some analysts and is highlighted here because it represents an analytical tool developed within a law enforcement context that may also be adapted for use in Homeland Security at the national level.

---

[142] Keith Harries, *Mapping Crime: Principle and Practice* (Washington, DC: U.S. Department of Justice, National Institute of Justice, Crime Mapping Research Center, 1999), 168

Geographic profiling is an investigative methodology that uses the locations of a connected series of crimes to determine the most probable area of offender residence. It is generally applied in cases of serial murder, rape, arson, and robbery, though it can be used in single crimes (auto theft, burglary, bombing) that involve multiple scenes or other significant geographic characteristics.

The basis of geographic profiling is the link between geographic crime site information and the known propensities of serial criminals in their selection of a target victim and location. The system produces a map of the most probable location of the criminal's centre of activity, which in most cases is the offender's residence. When linked with additional information relating to the crime incidents, and with additional data sources, such as motor vehicles databases and suspect databases, geographic profiling has been proven to have a profound impact on the effectiveness of a police investigation.

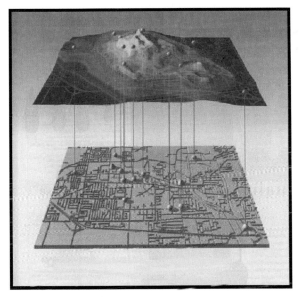

A geographic profile for the Abbotsford (British Columbia) Killer case

Source: Courtesy of D. Kim Rossmo.

Geographic profiling can be used as the basis for several investigative strategies, including suspect and tip prioritization, address-based searches of police record systems, patrol saturation and surveillance, neighbourhood canvasses and searches, DNA screening prioritization, Department of Motor Vehicle searches, postal/zip code prioritization and information request mail-outs. It is important to stress that geographic profiling does not solve cases, but rather provides a method for managing the large volume of information typically generated in major crime investigations. It should be regarded as one of several tools available to detectives, and is best employed in conjunction with other police methods. Geographic crime patterns are clues that, when properly decoded, can be used to point in the direction of the offender. [143]

Geographic profiling is a relatively new analytical tool that is best used by analysts trained in its methodology. Lorie Velarde, a crime analyst in Irvine, California (formerly in Garden Grove), has successfully used this tool in a number of situations, including burglary investigations. Pinellas County, Florida Sheriff's Department crime analysts also use geographic profiling as part of their repertoire of analytic techniques.

Geographic profiling has potential applications to homeland security, according to its originator, D. Kim Rossmo:

> Geographic profiling has a significant role to play in Homeland Security. Currently, a project by Texas State University with the U.S. Border Patrol is being completed on the "Geographic Patterns and Profiling of Illegal Crossings of the Southern U.S. Border," using environmental criminology principles to determine those physical and human geography factors associated with where the Texas/Mexico border is most porous. The goal of this project is to help the Border Patrol identify previously unknown areas where criminal border crossings may be occurring.

> Geographic profiling can also help focus investigative efforts on the planning operations of foreign terrorist cells within the United States, and on terrorist and insurgent activity directed against U.S. interests and coalition troops outside the United States. Finally, geographic profiling likely has a role to play in forensic epidemiology, including certain methods of bioterrorism.[144]

---

[143] From "What is geographic profiling?" (2001), at *http://www.geographicprofiling.com/geopro/index.html*.

[144] E-mail to the author from D. Kim Rossmo, 12 August 2005.

The SARA Model

The SARA model originates from problem-oriented policing's focus on *scanning, analysis, response,* and *assessment* as described on the chart below. Several analysts interviewed for this study emphasized the promise of building on this model of "what works" in law enforcement analysis.

The Problem-Solving Process and Evaluation

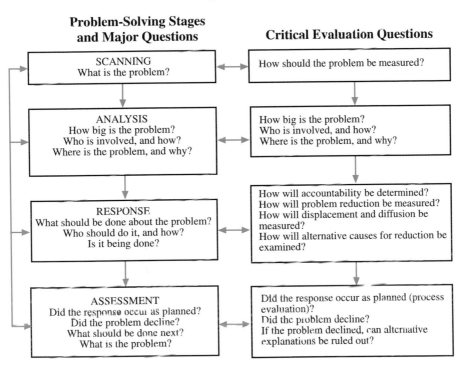

Problem-Solving Stages and Major Questions

Critical Evaluation Questions

| SCANNING — What is the problem? | How should the problem be measured? |

| ANALYSIS — How big is the problem? Who is involved, and how? Where is the problem, and why? | How big is the problem? Who is involved, and how? Where is the problem, and why? |

| RESPONSE — What should be done about the problem? Who should do it, and how? Is it being done? | How will accountability be determined? How will problem reduction be measured? How will displacement and diffusion be measured? How will alternative causes for reduction be examined? |

| ASSESSMENT — Did the response occur as planned? Did the problem decline? What should be done next? What is the problem? | Did the response occur as planned (process evaluation)? Did the problem decline? If the problem declined, can alternative explanations be ruled out? |

Source: John E. Eck, *Assessing Responses to Problems: An Introductory Guide for Police Problem-Solvers,* Center for Problem-Oriented Policing. Chart modified from URL: http://www.popcenter.org/Tools/tool-assessing.htm.

Problem Analysis

Problem analysis, discussed further in Chapter Ten, requires studying a crime problem in depth. Senior Public Safety Analyst Karin Schmerler, of the Chula Vista, California Police Department, has worked extensively in this area. The two charts that follow illustrate a few of the concepts behind problem analysis. The motel chart shows that budget motels, of all motels in Chula Vista, had disproportionate numbers of calls-for-service compared to other motels. Subsequently, Schmerler has been engaged in an in-depth study of the problem and, along with numerous other stakeholders, is working on strategies to reduce the cause of the problem at budget motels. The "top calls for service chart" below shows that many calls for police service are not crime related and implies that police resources might be more effectively used if those calls could be reduced in number by developing strategies specific to the problems.

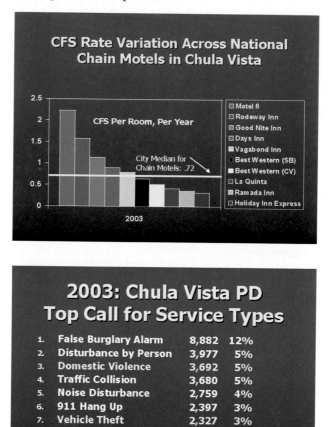

Source: Both charts courtesy of Karin Schmerler, Chula Vista PD.

Crime Series Identification and Predictive Modeling

A number of analysts indicate that Crime Series Identification and Predictive Modeling are valuable elements of their analytical work that should be carried forward to the future. Bryan Hill, of the Glendale, AZ Police Department, has developed techniques toward this end. He, like many other analysts, is adapting different methodologies to study and to solve the kinds of problems he faces in his daily work.

In my introductory stages of becoming a crime analyst, I was often confronted with the fact that probability ellipses and Gottlieb's rectangular probability areas were too large for practical use in many instances. Detectives and officers in our department often endow crime analysts with the mythical powers of the soothsayer of old. They are disappointed when you give them a 35-square mile area to patrol for the "next hit" by a robbery suspect. This is more often the case when a specific store type cannot be identified in the suspect's targets. By combining the efforts of several data sources and experience of the investigators and the crime analyst, a probability grid can easily be developed that will narrow the focus of the investigation. It can also assist the analyst in providing the investigating units a smaller group of businesses to put on their intensive patrol and stake-out lists. This is nothing new to geographers who often plan on where a new business would be built based on multiple layers of data being combined to come to the most logical location. In the Probability Grid Method, we use several different statistical methods to predict the next hit location, and do analysis within and between these layers to come to a final product that provides the best opportunities for the offender's next crime. [145]

Finding suspects in possible serial crimes is aided by mapping offender data, such as the next example, a map that shows incidents of violent crime in Knoxville, TN. In this case, as often happens, a clear pattern does not immediately emerge. Layering different sorts of data helps analysts examine possible relationships and the interpretation of such a map is only limited by the data available and the analyst's imagination.

---

[145] E-mail to author from Bryan Hill, 12 August 2005.

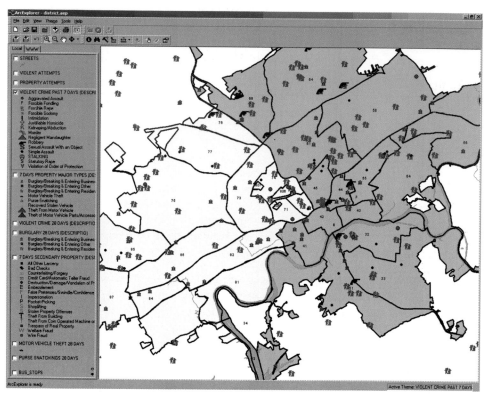

Mapping Crime — Knoxville, TN
Source: Lt. Robert Hubbs, Knoxville, TN, Police Department.

Next

Training, education, and partnerships with academia are elements of "things that work" for analysts and experts. Chapter Eight explores this subject.

Chapter 8

WHAT WORKS: TRAINING, EDUCATION, AND PARTNERSHIP WITH ACADEMIA

To those interviewed for this project, there is a distinct difference between training and education. Training is for the working analyst and involves learning how to use the tools and how to employ the techniques of the trade. It is vocational in nature. Training is frequently offered by private vendors, as well as by law enforcement associations and agencies. Education is a much broader class of learning, and several of the interview subjects have found that education in realms beyond the field of crime and intelligence analysis pays great dividends.

Training by Vendors

Anacapa Sciences, Inc., a company that innovated analytical training for law enforcement in 1971, remains a source of concepts for analytical techniques in law enforcement.[146] The Alpha Group Center is another company that has trained a number of law enforcement analysts.[147] Most law enforcement analysts had to rely on such private vendors[148] for training until recently. Now, with the growing recognition of the value of training for analysts, and the increasing number of analysts, a variety of programs has been developed.

Training by Associations

Both IALEIA and the IACA offer on-site training at their periodic conferences. IALEIA has contributed significantly to the development of the Foundations of Intelligence Analysis Training (FIAT) course. FIAT training provided by IALEIA[149] is analyst-instructed — many professionals hold that analysts should train analysts. "FIAT training was developed by a consortium of IALEIA, Regional Information Sharing System (RISS) Directors, the Law Enforcement Intelligence Unit (LEIU) and the National White Collar Crime Center (NWC3)."[150] The IACA has published *Exploring Crime Analysis*,[151] a textbook authored by analysts and academics. Chapters cover the twenty skill sets identified by IACA membership as essential to effective crime analysis.

---

[146] See *http://www.anacapasciences.com/company/overview.html.*

[147] See *http://www.alphagroupcenter.com/About_AGC.html.*

[148] Information on scheduled training can be found at *www.ialeia.org and www.iaca.net.*

[149] Further information on IALEIA FIAT training information can be found at *http://www.ialeia.org/training.html.*

[150] See *http://www.ialeia.org/protocol.html.*

[151] *Exploring Crime Analysis:* Readings on Essential Skills (North Charleston, South Carolina: Booksurge, LLC, 2004). URL *http://www.iaca.net/ExploringCA.asp.*

WHAT WORKS

FLEAA

The Florida Law Enforcement Analyst Academy (FLEAA) is the first of its kind. The Florida Department of Law Enforcement (FDLE) recognized the critical need for analyst training at the local level of law enforcement and invested Homeland Security funds into its creation and continuation. As of June 2005, four classes have graduated ninety-seven analysts from over fifty municipalities, sheriff's offices and state agencies.

The new war on terrorism has placed a premium on a higher level of analysis, with the primary function of predicting and preventing criminal activity. Funded through the U.S. Department of Homeland Security (DHS) to advance the capabilities of law enforcement agencies throughout the state, FLEAA is a highly specialized training program for crime intelligence analysts assigned to anti-terror investigations. Available to the entire Florida law enforcement community, FLEAA is a standardized advanced training curriculum in intelligence analysis, data management, and crime pattern analysis.[152]

FDLE offers a definition of "law enforcement analyst":

"Law Enforcement Analyst" means any person who is employed or contracted by any municipality or the state or any political subdivision thereof; whose primary responsibility is to collect, analyze and disseminate data to support, enhance, and direct law enforcement missions. Their efforts are directed at identifying criminal subjects, organizations, activities, events and/or forecasting future crime occurrences utilizing analytical techniques. This definition includes all law enforcement analysts who provide strategic, operational, investigative, intelligence and crime analysis.[153]

FLEEA has improved the capabilities of law enforcement analysts and other emergency responders to develop and use information and intelligence to prevent crime and conduct complex investigations. The establishment of this academy also puts in place a career path in investigations for non-sworn law enforcement personnel. "The curriculum for the academy covers a wide range of topics inclusive of the many analyst functions in the law enforcement agencies, including crime analysis, intelligence analysis, investigative analysis, and management/statistical analysis."[154]

Strategic Assessment: An Example from an Analyst Academy Graduate

The Analyst Academy requires analysts to complete a strategic assessment as a condition of graduation. The analyst and decisionmaker in a given agency determine the subject

---

[152] See *http://www.it.ojp.gov/documents/20050615_FDLE_FLEAA_graduates.pdf.*

[153] Courtesy of Lois Higgins, program administrator.

[154] From MS PowerPoint slides provided by Lois Higgins.

of the strategic assessment. A real need is identified and addressed through the assessment, and, in this way, the command staff of the respective agency is introduced to the concept of using analysts to understand problems in greater depth as well as to identify solutions or other types of recommendations. An example:

> The purpose of this assessment was to review several pieces of data to determine if there is any correlation between the "cycles" associated with heroin abuse and the number of burglaries committed. The goal is to explore the possibility of developing strategies that will reduce the number of burglaries committed in the City of Kissimmee by focusing on the heroin abuser.
>
> Peak times in which burglaries are being committed citywide as well as those in and around documented "open air drug markets" were analyzed. Information from narcotics investigators in relation to heroin abuse was also analyzed. Drug and burglary arrests data, peak times for narcotics related calls, and geographic information with respect to where known heroin abusers are living was collected and analyzed.
>
> The results of this assessment will be a better understanding of the burglary problem in the City of Kissimmee, heroin abuse, and the possible connection between the two.[155]

Few analysts in local law enforcement work in the capacity of investigative support. Analysts at this organizational level have traditionally focused on administrative reporting and tactical support. The training provided in Florida will help standardize the work of analysts across that state's agencies, enabling specialists in various areas to speak the same language when they think of the word "analyst." The FDLE does recognize the transcendent value that professional analysts can bring to cognate fields of law enforcement.

> The value of competent and professional law enforcement analysts in support of investigations continues to become even more evident. The anticipated impact of intelligence-certified analysts is enhanced investigative outcomes and professionalism in this critical investigative support position. [156]

---

[155]Strategic assessment overview provided by Crime Analyst Metre L. Lewis, Kissimmee Police Department.
[156]See URL *http://www.it.ojp.gov/documents/20050615_FDLE_FLEAA_graduates.pdf.*

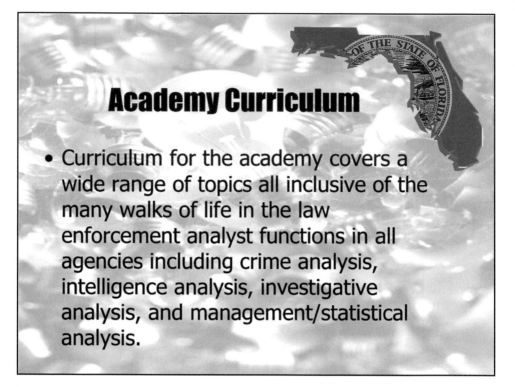

Academy Curriculum

- Curriculum for the academy covers a wide range of topics all inclusive of the many walks of life in the law enforcement analyst functions in all agencies including crime analysis, intelligence analysis, investigative analysis, and management/statistical analysis.

Source: Lois Higgins, Florida Department of Law Enforcement, FLEAA.

The Counterdrug Intelligence Executive Secretariat

The Counterdrug Intelligence Secretariat of the Office of National Drug Control Policy published a Basic Training Curriculum for Law Enforcement Analysts in March 2003. It is not widely available; however; it was used in the development of the FDLEAA. One of the attention-getting assertions by the Secretariat:

Analysts' Job Focus: Professionalizing the intelligence analytic cadre at Federal law enforcement agencies … requires that intelligence analysts will no longer perform data-entry tasks and other nonanalytic-related tasks such as technical or graphics support, but rather focus on the job for which they were hired — research and analysis.[157]

The International Association of Law Enforcement Analysts and other state, regional and federal law-enforcement agency representatives did have input into developing this training curriculum. Since the FLEAA addresses all law enforcement

---

[157]See *http://www.whitehousedrugpolicy.gov/publications/gcip/sectione.html,* 2002, paragraph E-19.

enterprises in Florida, and many academy graduates function as municipal crime analysts, the adapted curriculum had to include their training needs as well as those proposed by the Counterdrug Intelligence Secretariat for this curriculum. Ms. Higgins emphasized to the author the direct linkage that appears to exist between the analytical techniques that were adopted into this curriculum and the prospects for receiving approval for funding. As a state-level analyst, she recognizes the value of the curriculum, but she also contends that the mission of local-level law enforcement requires an emphasis on different methods of analysis.

A Model for Training: Police Service of Northern Ireland's Analysis Centre

Reforms in policing in Northern Ireland have led to the development of one of the most progressive analytical capacities existing in law enforcement today.

The Independent Commission on Policing in Northern Ireland was set up as part of the Agreement reached in Belfast on 10 April 1998. The task of the Commission was to provide "a new beginning to policing" in Northern Ireland. In its report published in 1999 [commonly referred to as the Patten Report], the Commission made 175 recommendations about policing in Northern Ireland. Amongst the recommendations were proposals regarding the composition, size and structure of the Police Service. It also recommended the creation of new accountability structures, and said that Human Rights and community policing should underline all of the work carried out by the Police Service.[158]

The PSNI has developed a comprehensive training program for law enforcement analysts in an Analysis Centre. With the advent of Northern Ireland's policing reforms late in the 1990s, funding and political will grew sufficient to restructure the business of policing with a focus on the intelligent use of analytical support. Created in December 1999 by the Chief Constables Policy Group, the Analysis Centre strives to be "a Centre of Excellence in the professional development and delivery of analytical services and products. Police analysts in Northern Ireland work to national standards." [159]

The PSNI was involved with representatives from European Union Member states experienced in the fields of intelligence and analysis in the production of a booklet entitled *Intelligence Management Model for Europe* (phase 1) — Guidelines and standards and best practice within the analysis function.[160]

Mark Evans, Director of the Centre, indicates that in late 2005, details about the "Analysis Centre Model" will be incorporated into one or more reference manuals. [161]

---

[158]See *http://www.psni.police.uk/index/about_psni.htm.*
[159]See *http://www.psni.police.uk/index/departments/analysis_centre.htm.*
[160]See *http://www.psni.police.uk/index/departments/analysis_centre/*
pg_intelligence_management_model.htm.
[161]E-mail from Mark Evans to author, 24 July 2005.

Analysts from the PSNI have come to North America "to exchange best practices ideas and techniques."[162] Participant Lt. Robert Hubbs of the Knoxville Police Department reports that hosting a PSNI analyst was one of the high points in his work as an officer. More detailed information regarding the PSNI Analysis Centre may be found at *http:// www.psni.police.uk/index/departments/analysis_centre.htm*. This Center, in the view of the present author, is the most effective example now available for developing the analytical capacity of local law enforcement and integrating it with other levels of law enforcement, government, and the community.

Analysts' Training in the U.S. National Intelligence Community (IC)

Training for analysts remains a contentious issue in the IC.

Until very recently, the intelligence community did not spend a significant amount of time on analyst training. This training is most useful in giving in-coming analysts a sense of how the larger community works, what is expected of them, and the ethos and rules of the community. No amount of training, however, can obviate the fact that much of what an analyst learns comes through on-the-job training.[163]

Yet, in many law enforcement agencies, there is no one even to provide on-the-job training for analysts. They depend instead on outside resources to learn their work — a situation that has distinct limitations.

Education

Some new academic programs focus on crime and intelligence analysis, but they are slow to develop, when we consider how important analysis is to the intelligence process. This observation suggests that validation is lacking for the concept. If there are few texts and little university study of an otherwise important concept, then for important sectors in our society, the concept has little basis in "reality."

Several subjects interviewed for the present study are teaching college courses on crime and/or intelligence analysis. The Research and Intelligence Analyst Program (RIAP) at Mercyhurst College focuses on teaching analytical skills for federal and national levels of government, as well as on business intelligence. Tiffin University is offering an online Master's program with a concentration in crime analysis. The California Department of Justice sponsors a credentialing program in four California universities that is focused on law enforcement crime and intelligence analysis; all students who successfully complete the credential program become "Certified Crime Analysts." [164]

---

[162] See URL *http://www.psni.police.uk/index/departments/analysis_centre/ pg_international_analyst_exchange_programme.htm.*

[163]Lowenthal, *Intelligence*, 90.

[164] California State University, Fullerton; California State University, Northridge; California State University, Sacramento; and the University of California, Riverside are the California universities offering this program of study, which is also affiliated with the Alpha Group Center. See *http:// www.alphagroupcenter.com/Certification_Info.html.*

Still, in exchanges with the author, experts most often suggested that a broad education in the social sciences and the humanities, with an emphasis on research, is preferred for budding law enforcement intelligence analysts. A broadly educated and strongly motivated individual with an ability to learn is more likely to be a good analyst than someone who does not have these intrinsic traits.

Law Enforcement Partnerships with Academia

Many of the participants in this study have worked with academic professionals. Some are themselves academics who have been influential in shaping the face of policing today. Several started as police officers who went on to be significant contributors to innovation in law enforcement analysis through their grounded research.

From conversations with the experts, the author found that academics tend to partner more frequently with local law enforcement officials in active, practical research than with other levels of law enforcement. This observation may not withstand more systematic inquiry into the phenomenon, but it seems that large-scale, regional or interstate investigations are usually so restricted to those with the "need to know" that they would not be open to having "outside" eyes and ears examining potential areas of improvement and innovation.

A spotlight can also be shone on other challenges to potential partnerships between law enforcement officials and academics in the conduct of applied research:

The reality of law enforcement agencies and research departments is that their institutional cultures and requirements may, and do, lead them towards different research agendas. Law enforcement agencies are typically interested in research focused on local, daily policing practices. They must pursue research that responds to external political realities. This need to address political pressures similarly dictates that law enforcement agencies require qualitative as well as quantitative research in relatively compressed time frames. Law enforcement leaders work in a world where public opinion matters and must be understood and addressed.

The interests of university-based research departments tend to [diverge] systematically from those of law enforcement agencies, and researchers shape their desired research agendas accordingly. Frequently, faculty in pursuit of tenure or students in pursuit of faculty positions value research that addresses national concerns [and] that may ultimately change the face of policing, but [such] broad-brush efforts … are of only limited interest to local law enforcement agencies. Similarly, researchers may not view the political pressures that influence police chiefs as being as important as those political pressures that shape the interests of university departments. They may also value quantitative research over qualitative research.

While institutional priorities differ, and may frequently result in different research priorities, the importance of research results dictates that neither law enforcement agencies nor research departments can afford to allow such differences to hinder the establishment of research partnerships. While a basic familiarity with each other's institutional priorities will always be important, a specific awareness of each other's research interests is vital to the establishment of productive partnerships.[165]

This IACP report goes on to encourage greater mutual understanding between the participants in research partnerships:

> Law enforcement agencies and research departments should recognize that the best research will be characterized both by its relevance to policing practices and its theoretical sophistication. Research departments, in particular, should work to validate action research vital to improving policing practices.[166]

A number of analysts in law enforcement are "real" researchers. However, the IACP report does not recognize the potential benefit of networking analysts with outside researchers to develop more relevant knowledge and tools. This linkage is very common in other fields, and is a valuable strategy. Analysts tend to be better educated than many of their agency counterparts. "I am a blue-collar academic demonstrating the utility of theory, techniques, and statistics," claims one analyst. They understand the needs of their agencies and the needs of academics. They can and do, in fact, "translate" and facilitate between the two camps.

The dichotomy between the roles of analyst and academic is, nonetheless, an artificial one. We need academics to be employed and integrated into law enforcement to develop a cadre of knowledge workers. The paradigm of having outside experts who allegedly "know better" can be changed by changing hiring practices within law enforcement, as a means of championing the professionalization of this applied field.

An Example of Partnership with Academia: Operation CEASE FIRE

On 31 March and 1 April 2005, the author attended an Operation CEASE FIRE conference in Rochester, New York. Subsequently, Chris Delaney, a research student affiliated with the Rochester Institute of Technology, who worked in Rochester on its CEASE FIRE intervention, was interviewed for this project. The operation involves gathering a great deal of intelligence about violent gang and street criminal groups. A target group is identified and all its known members fully prosecuted after a violent incident, usually a homicide. They, in turn, are used as an example for other gangs, or for those already serving time in the criminal justice system, or on parole or probation. The gang/group members are warned that if there is another homicide and the police have identified who was

---

[165]National Institute of Justice (NIJ)/International Association of Chiefs of Police (IACP), *Unresolved Problems and Powerful Potentials: Improving Partnerships Between Law Enforcement Leaders and University Based Researchers: Recommendations from the IACP 2003 Roundtable*, August 2004, 14-15. Read the full report at *http://www.theiacp.org/documents/pdfs/Publications/LawEnforcement%2DUniversityPartnership%2Epdf.*

[166]*Unresolved Problems and Powerful Potentials*, 16.

involved, they and every member of their group will be targeted for full enforcement/ prosecution.

The Boston Strategy to Reduce Youth Violence,[167]of which Operation CEASE FIRE was a main component, involved the collaboration of many stakeholders. Within the strategy, academia played a role in intelligence gathering and analysis in a way seldom seen. This strategy proved effective in Boston. David Kennedy, the Harvard researcher (now at John Jay College in New York) presently is at work in New York State to implement this strategy with multiple law enforcement agencies. The collection of intelligence from street-level officers drives the strategy employed in these operations. Academics help organize the gathering, collection, and updating of intelligence, including the production of gang territory maps, as described here:

> We looked at the criminal histories of five years of youth victims and offenders and found very high rates of prior arrests and court involvement. We looked at where gangs identified turf. We built gang maps and network maps of gang relationships based on the professional knowledge of practitioners. This was not available from old documents and paper records, but when you put these very skilled, experienced front-line folks around a table and a flip chart and got systematic about what they knew, they knew all this. And we ended up digitizing these maps and playing all kinds of fancy games with them, but it was all built on what they knew and the work they had done.[168]

---

[167] Read details about this program at *http://www.bostonstrategy.com/programs/ 11_OpCeaseFire.html.*

[168] David Kennedy at *http://www.bostonstrategy.com/players/04_academia/01_academia.html.*

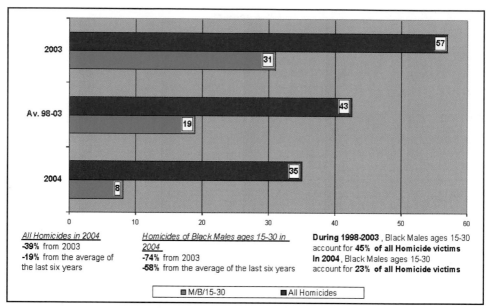

Chart data labels:

- 2003: 57 (All Homicides), 31 (M/B/15-30)
- Av. 98-03: 43 (All Homicides), 19 (M/B/15-30)
- 2004: 35 (All Homicides), 8 (M/B/15-30)

All Homicides in 2004
-39% from 2003
-19% from the average of the last six years

Homicides of Black Males ages 15-30 in 2004
-74% from 2003
-58% from the average of the last six years

During 1998-2003, Black Males ages 15-30 account for **45% of all Homicide victims**
In 2004, Black Males ages 15-30 account for **23% of all Homicide victims**

■ M/B/15-30   ■ All Homicides

Effect of operation cease fire in Rochester, NY

Source: Chris Delaney, Rochester, New York Police Department.

The chart above demonstrates the effectiveness of the strategy in Rochester. Partnerships with academia can result in real-world progress and are valuable to the professional education and experience of many analysts.

Next

The nature of each particular organization itself plays an important role in the success of a law enforcement analyst. Institutional policing practices also impact on success, or lack of success. The following chapter explores what works in law enforcement organizations and institutions from the analyst's perspective.

Chapter 9

AN ORGANIZATIONAL AND INSTITUTIONAL VIEW
OF WHAT WORKS

Agencies have to empower analysts to be analysts.[169]

Analysts are often those who develop a lead, identify a connection between affiliates, or determine the location of a suspect. The role of analysis in a law enforcement agency is to support the investigative, intelligence, and planning activities of the agency. [170]

We wander around until someone calls us for help. [171]

Tradition versus Progress

Limited resources and fragmentation of services impair the movement of law enforcement from reactive modes to proactive modes. For law enforcement analysis to "work," to help law enforcement be proactive in using information, law enforcement agencies must understand the role of the analysts and their work must be valued.

The nation's police departments, the majority of which have fewer than 50 sworn officers, continue traditional operations despite a rapidly changing social and scientific landscape. Working in small communities with limited and overlapping jurisdictions and dependent upon austere local budgets, most police departments struggle to provide even the most basic traditional investigative and uniformed-response services. Computer technology for a large percentage of the nation's smaller police departments is extremely limited. [172]

Analysts have no clear sense of what products are useful to the target [consumer]. They produce, deliver, and through anecdotal evidence draw conclusions about the value of their work. Neither the analysts nor the analysts' managers are clear about how and how well the targets use their products. Individual quantitative and qualitative measures cannot be established until managers know what products crime analysis units should be producing. This must come from a systematic assessment of crime analysis output. [173]

Interview subjects expressed widespread frustration over organizational and developmental obstacles to the application of crime and intelligence analysis in law enforcement. One of their fondest wishes for the future is that law enforcement managers understand

---

[169] Quote from interview.
[170] FDLE press release. URL *http://www.it.ojp.gov/documents 20050615_FDLE_FLEAA_graduates.pdf.*
[171] Quote from interview.
[172] Cowper and Buerger, 10.
[173] O'Shea, *Crime Analysis in America, 16.*

the role of information analysis, its value, and its limitations. Some analysts believe it is necessary for them to rise through the ranks and become supervisors and policymakers in order to change the status quo and increase analytical effectiveness.

Crime Patterns and Organizational Responses

Perhaps the most important foundation for success in law enforcement intelligence analysis, as seen by the subjects involved in this study, is embodied by an organization that acts on analytical information, especially in relation to crime pattern identification and predictive crime series reports. More than one analyst reported making accurate predictions regarding when and where a crime will occur, but resources and/or organizational responses were not provided to stop the crime and/or apprehend the criminal. In the author's experience as well, organizations that are structured to act on various sorts of information — to respond and problem-solve — are the *sine qua non* of analytical success. In British terms:

Incidents may be patterned in numerous ways, giving clues to potential problem-solving interventions at levels below the individual incident. There is now ample evidence that incidents are good predictors of future incidents, especially in the short term. This provides a focus for problem-solving at the level of service delivery. Incidents are also patterned at wider levels, providing further opportunities for problem-solving. They can be patterned by time (of day, day of the week, time of year, time of moving house etc); by place (nearness to junction, street, street side, part of town etc); by victim attribute (sex, age, ethnicity, organisation type, income, type of residence, household type, security levels etc); by *modus operandi* (victim route to offence, escape from offence, mode of and direction of entry, type of attack, goods stolen, means of disposal etc); or by offender attribute (education, sex, age, co-offending practices, network of associates, form of organisation, lifestyle, drug and alcohol taking habits etc). They can also, of course, be patterned by mixes of these sorts of attributes. Different crimes will show different patterns in different places. Informed and imaginative analysts, supplied with good data from a range of sources, will be needed to tease out patterns identifying problem-solving needs and opportunities. Preventive problem-solving opportunities emerge from identifying, analysing, and anticipating patterns. Differing levels and types of organisation can problem-solve at differing levels.[174]

---

[174] Tim Read and Nick Tilley, *Not Rocket Science? Problem-Solving and Crime Reduction.* Crime Reduction Research Series Paper 6 (London, Home Office: Policing and Reducing Crime Unit Research, Development and Statistics Directorate, 2000), 3-398. At URL *http://www.homeoffice.gov.uk/rds/prgpdfs/crrs06.pdf.*

WHAT WORKS

Leadership

Supportive, responsive, and understanding leadership are *necessary* for analysts to be most effective. The leadership of an organization determines its priorities. To become intelligence-led, a law enforcement agency's leadership must be dedicated to analysis not only in word, but in action — this means adequate resources for analysis, including ongoing training for staff and relevant and effective responses to analytical output. Together, dedicated leadership and analyst professionalism make an organization *able* to live up to the ideal described here:

> There is nothing more important to an individual committed to his or her personal growth than a supportive environment. An organization committed to personal mastery can provide that environment by continually encouraging personal vision, commitment to the truth, and a willingness to face honestly the gaps between the two.[175]

A Leadership Example from Lincoln, Nebraska

Chief Tom Casady of the Lincoln, Nebraska PD empowers his officers to do analytical work and find the information they need by providing IT systems that are easy to navigate, and training officers in their efficient use. Chief Casady goes a step beyond this by testing officers for analytical skills as a requirement for promotion. Twenty percent of the areas tested for promotion involve analytical skills. Examples of test questions are in the text boxes below.

---

[175]Senge, 173.

Lincoln, Nebraska Police Department: Example of Promotion Test

Information Retrieval Exercises:

There are seven exercises, some with multiple questions. You must work independently and without assistance from others. You may use the Alpha terminal and PC at your workstation. Write your responses in the space provided. If you wish, you may print any material and attach it as part of your response.

You have a total of 30 minutes.

Exercise 1

A caller is displeased with the police response to repeated disturbance complaints at what he asserts is a "party house" at 3243 N. 49th.

- How many complaints of this type have we handled in the previous 12 months?
- Who has been complaining?
- Who owns the property?
- Who has been cited, for what, and what has been the outcome of any citations?

Exercise 2

List the names and addresses of the Juvenile Drug Court defendants who reside on the Southwest Team area.

Exercise 3

List the date, case number, and violation type for each of the tavern violations reported at The Downtown during the previous 12 months.

Exercise 4

This morning, a purse snatch occurred in which the victim was attacked just south of the Capitol while walking to work around 0700 hours. Two citizens chased down the suspect, and an arrest has been made. This case sounds vaguely familiar to you, and you recall another purse snatching in the area directly south of the downtown area within the past few weeks. It stood out in your mind because of the early hour. Find the case number of that case. What is the description of the suspect in that case?

Exercise 5

The president of the Forecaster Elementary School PTA has asked you to attend their meeting tonight. He wants you to provide him with some information about police calls or incidents in the general area of the school so far this month. You are unable to attend due to a prior commitment, but intend to send your second shift sergeant.

Provide a synopsis of this activity for your sergeant to take along.

Exercise 6

The clerk of the City Council has notified the chief's office that a citizen intends to come to the City Council meeting this afternoon during the open microphone session to complain about the lack of traffic enforcement at Cotner and Sumner. She lives on the corner, and says that there have been several bad accidents at the intersection because people are always running the stop signs. The chief needs the accident history of this intersection since January 1, 2003. Since the chief will not be back at the office prior to the meeting, he wants you to send the information to him at his personal email address, so he can retrieve it on his cellphone. Write the information down in the space below, but also email your response.

Exercise 7

During the investigation of some vandalism cases, a golf club was recovered that was used to smash car windows. The suspect in the vandalism claims he found the golf club lying on the ground near Eastridge swimming pool. The club is a wedge inscribed with "Greg Wilson" on the shaft. You think it's probably stolen. Where did it come from?

Source: Courtesy of Chief Tom Casady, Lincoln, Nebraska Police Department.

All employees at all levels are empowered to use information in his organization. He has removed the proprietary nature of intelligence information to make it more available rather than less available. He states, "The protection of intelligence information will restrict it so that it will lose its value." [176] He would rather risk sharing it to obtain its value than not.

In Lincoln, crime analysts and intelligence analysts are co-located and work on higher levels of analytical tasks than those expected of sworn officers. Officers in Lincoln can on their own volition get lists of crimes or bar charts showing crime trends, using the interface screen shown on the next page. The crime analysts are thus freed up to do special projects, such as completing detailed dossiers on people who have committed gun crimes in the last thirty days. They would do this in collaboration with the regional U.S. Attorney's office.

---

[176] Interview, 6 January 2005.

Source: Courtesy of Chief Tom Casady, Lincoln, NE Police Department.

Elevation of the Analysts' Role

Analysts working on the same bureaucratic level with command staff serve to facilitate integration of intelligence work into a law enforcement agency. This approach remains uncommon at this time, but it is a direction that analysts sampled in this study would propose to help in the development of analysis in law enforcement. The following vignette describes an example of such a process underway in the RCMP (Royal Canadian Mounted Police):

> The RCMP has recognized that the effectiveness of its intelligence programme is completely dependent on its personnel and has taken a number of innovative steps related to analysts' positions within the organization. Civilian member analysts are now paid a salary comparable to that earned by their regular member counterparts. Beyond this, their value to the organization is reflected in their rank. The civilian manager of the overall analytical programme is a Superintendent-equivalent, the supervisors of the various analytical sections are Inspector-equivalents, while the analysts themselves occupy a range of NCO (Corporal/Sergeant/Staff Sergeant)-equivalent positions.

Focusing on Repeat Offenders

If, as D. Kim Rossmo asserts in Geographic Profiling, 10 percent of offenders commit 50 percent of crime, an effective strategy is for agencies to focus on identifying and more closely monitoring those 10 percent in their jurisdictions (and in surrounding areas) who consistently re-offend. Analytical effectiveness is enhanced if analysts are used to help identify those among the 10 percent, analyze and report on the qualitative aspects of their behavior, and recommend strategies to either apprehend them the next time they re-offend or prevent their crimes. This type of information must be continually updated. Analysts also engage in comparative case analysis to help officers demonstrate to prosecutors that the offender is responsible for more than one crime.

> Much of the crime we fear most is committed by repeat offenders. Many of the nation's investigations, arrests, prosecutions, and criminal trials are devoted to chronic offenders who are out on bail, on felony probation, on parole, or on early release into the community because of prison overcrowding. These relatively few offenders impose an immense burden on the criminal justice system. To prevail in its mission, the system must work "smarter, not necessarily harder," by removing repeat offenders from the community. [177]

Focusing on repeat offenders must be an organizational priority if it is to be effective. It requires sharing intelligence with the entire organization. This threatens some officers who would prefer to keep information to use for their own arrests. This may also threaten the advent of investigations by an outside agency. Yet, if we have a chronic offender who is unknown to most officers but known to a few, the prospective intelligence failure is clearly inane — and preventable.

Modernity

Organizations that embrace technology and adapt to changes in technology facilitate the success of analysis. One such example is illustrated below.

For over four years the Knoxville, Tennessee, Police Department has been using GIS software for problem-oriented policing, crime mapping, and deployment analysis. The Knoxville Police Department serves over 175,000 people and is staffed with over 380 police officers. The Arc Explorer interface displays 15 data layers and a color digital photo for pattern/trend detection, for directed patrol, and for strategic and tactical planning. This includes violent, property and sex crimes, calls for service, and traffic collision locations. Data specific to the incident are displayed, and using the map tips feature

---

[177]Susan E. Martin and Lawrence W. Sherman, *Catching Career Criminals: The Washington, DC Repeat Offender Project,* Police Foundation Report (Washington, DC: U.S. DOJ, National Institute of Justice, 1986), 5.

Mapping in Patrol Cars: Knoxville, Tennessee
Source: Courtesy of Lt. Robert Hubbs, Knoxville, TN, Police Department.

allows officers to pass the map pointer over each point and see date, time, location, and other details in a pop-up label.

Map data are distributed daily throughout the city using a wide-area wireless network. There are over 20 wireless sites throughout the 100 square miles of Knoxville city limits. Officers, using the 250 field laptops, can go to any of these areas and quickly download the latest map data. Seven- and 28-day trend maps are displayed along with the mapped addresses of parolees, probationers, and sex offenders. The school system supplies bus stop locations that are compared to the work and home addresses of sex offenders for risk assessments.

Officers can see 90 days of shots-fired calls, and related gun crimes, allowing officers to develop strategies of interdiction and tactical deployment. The field mapping initiative assists Project Safe Neighborhoods, a nationwide commitment to reduce gun crime in America that is led by the United States Attorney in all of the 94 federal judicial districts across the country. The system networks existing local programs that target gun crime and provides those programs with additional tools necessary to be successful. The goal is to create safer neighborhoods by reducing gun violence and sustaining the reduction.[178]

Educated Managers and Coworkers

Educated law enforcement managers and educated law enforcement officers are a key to the success of analysis in law enforcement. Few training resources exist to educate law

---

[178] Lt. Robert Hubbs, Knoxville, TN PD.

enforcement managers. One innovative example of training for managers is offered by *The Intelligence Centre*, an Australian-based vendor. One of the interview subjects, Howard Clarke, teaches this course.

The focus of the training is centred on the following three elements:

- the intelligence process, role and functions of the analyst;
- the intelligence supervisor — including managing the intelligence activity;
- the client — including tasking, monitoring and assessing performance of intelligence services.

The workshops discuss the concepts and practice of intelligence and provide several opportunities for participants to practice and role-play in the three roles of analyst, supervisor and client.

Workshops may be conducted separately for supervisors or managers or clients. However, mixed groups generally provide better opportunities for more effective understanding of different needs and perspectives.[179]

This type of training is critical to the development of analysis in law enforcement. Most subjects expressed frustration at the lack of understanding law enforcement managers have about their role, and the difficulty in obtaining the information and tools to do their jobs effectively. One analyst said that this resulted in a positive: "forced creativity," or the need to find ways to work around the obstacles presented by being tasked to do this job without easy assess to the resources needed. Another analyst stressed the need for a better-educated intelligence consumer, one who knows what intelligence can and cannot do. He referred to a pronouncement thought to be associated with Henry Kissinger, "I don't know what kind of intelligence I need," which suggests that he was open to learning about intelligence capabilities.

Bioterrorism, identity theft, cyberstalking, and crimes not yet defined will require more intelligent, better educated and trained, and more tech-savvy officers and leaders than are now available in policing. Today a large majority of agencies require only a high school diploma or equivalent to begin a career in policing. Training usually consists of 12 to 16 weeks of academy work and a six-month probationary period, during which superiors evaluate the new officer. A few departments have specialists coping with Internet crime, but most do not. Leadership and management courses are offered for some, but coping with technology and transnational crime has just begun to be part of the training.[180]

---

[179] See *http://www.intstudycen.com/Training_Programs/Manager_W_Shop/ manager_w_shop.html*. Interview subject Howard Clarke, contract analyst for the RCMP, teaches for the Intelligence Centre. A key textbook from the Intelligence Study Centre is Don McDowell, *Strategic Intelligence: A Handbook for Practitioners, Managers and Users* (Cooma, NSW, Australia: Istana Enterprises Pty, Ltd, 1998).

[180] Gene Stephens, "Policing the Future: Law Enforcement's New Challenges," *The Futurist* 39, no. 2 (March-April 2005), 57.

Educated officers also enhance the utility of analytical products.

Advances in information processing, and hopefully advances in data analysis methods, will provide the police with unprecedented opportunities to broaden their mission and mandate in a manner that may well impact the nature and extent of crime, not merely react to its occurrence. This will require police policymakers to critically assess their current operations and management paradigm. They will have to permit themselves to be open to new ideas, methods, and practices and provide employees within the department with the necessary skills, resources, and time to conduct such problem analytic tasks. It will not be enough for the academics, or anyone for that matter to provide the tools and techniques; police managers will have to understand and appreciate the value of these new tools and techniques and demand the products that they imply.[181]

Academia and Police Managers

Research dollars spent helping academics develop partnerships with police have brought excellent results. Yet, until police managers invest in a more scientific approach to analysis and problem solving, their efforts will bear fruit only in places fortunate enough to have insightful, progressive police managers who will take ownership of the possible applications. A recent, federally funded study notes that:

For the past fifteen years academics have been given unprecedented access to police data. Partnerships have flourished between police and academics. Yet, when it comes to practical applications of data analysis to inform problem analysis, academics have not been up to the challenge. It is one thing to say that the police should focus their attention on understanding the factors behind patterns and relationships in the police data to better explain the complexities of the criminal and disorder. It is quite another thing to show how.

We would hope that contemporary police managers can bring themselves to think outside the box. Advances in information processing, and hopefully advances in data analysis methods, will provide the police with unprecedented opportunities to broaden their mission and mandate in a manner that may well impact the nature and extent of crime, not merely react to its occurrence. This will require police policymakers to critically assess their current operations and management paradigm. They will have to permit themselves to be open to new ideas, methods, and practices and provide employees within the department with the necessary skills, resources and time to conduct such problem analytic tasks. It will not be enough for the academics, or anyone for that matter, to provide the tools and techniques; police managers

---

[181] Timothy C. O'Shea, Crime Analysis in America: Findings and Recommendations (U.S. DOJ, Office of Community Oriented Policing Services, 2003), 26. URL *http://www.cops.usdoj.gov/mime/open.pdf?Item=855.*

will have to understand and appreciate the value of these new tools and techniques and demand the products that they imply.[182]

Next

The next chapter explores the future of law enforcement analysis.

---

[182] O'Shea, 26.

Chapter 10

THE REAL AND THE IMAGINED FUTURE

Imagination is not a gift usually associated with bureaucracies.[183]

Use intellect, information, and technology to enable and empower preventable measures that make our community safer.[184]

Since the number of participants who can join a partnership is infinite, so are the options we can devise to prevent crime.[185]

...computer chips, sensors, software and actuators can now turn any kind of passive product into an active preventer, and possibly even an intelligent one. The day is here when we can make a reality of the magic harp in Jack in the Beanstalk — one that doesn't only cry "Master, Master, he's stealing me!" but also sends the Giant a warning by email on his WAP capable phone, and shuts down, refusing to play until he's back in the Giant's castle or there's been a formal transfer of ownership.[186]

Paradigm Shifts

Some ideas suggested by the interview subjects reflect wishes for the future. Some of these ideas are already realities in the field of law enforcement, even though they are not widespread in application. The next sections introduce a few such concepts: a new intelligence cycle, problem policing, crime science, and neighborhood involvement.

A New Intelligence Cycle: Interpret, Influence, Impact

Jerry Ratcliffe articulated an alternative view of the purpose of intelligence and crime analysis in a world that is evolving toward intelligence-led policing. His vision is more holistic than the standard intelligence cycle model; it includes the desired consequence of law enforcement analysis — to make an impact on crime. Although getting specific criminals off the street has great virtue, if there is no reduction in crime, we must ask ourselves: How effective is our work? Where is the meaning if our work does not effect the changes we say constitute the core of our mission to protect and to serve?

A few years ago, I was looking for a simple model to explain to intelligence and crime analysts what their job should be in an intelligence-led policing environment.

---

[183] *9/11 Commission Report*, Foresight-Hindsight, 6, online version at URL *http://www.9-11commission.gov/*.

[184] Quote from interview.

[185] Debra Cohen, *Problem-Solving Partnerships: Including the Community for a Change* (USDOJ: COPS. June 2001), from Moscow [Idaho] Police Department PERF Response/Assessment Survey, 1999:12.

[186] Paul Ekblomb, "Less crime by design," RSA Lectures, 10 November 2000. URL *http://www.rsa.org.uk/acrobat/Ekblomb.pdf*.

The problem was that many analysts spent their whole time analysing crime and not thinking about the broader picture: why they were doing the analysis, and what they should do with the results. The 3-I model grew from discussions with an Australian federal agent and aims to show the purpose of crime analysis in the modern policing world. The 3-I model revolves around three law enforcement activities:

- Interpret
- Influence
- Impact [187]

The 3-I Model for Police Intelligence

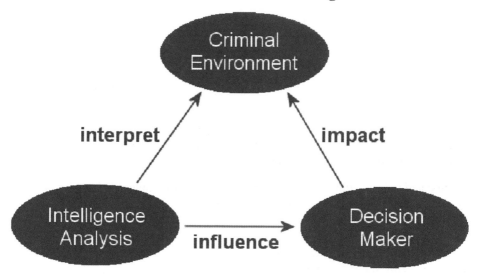

Source: Jerry H. Ratcliffe, URL *http://jratcliffe.net/research/three_i_model.htm.*

For the future of policing as well as homeland security, creating intelligence products that contribute to the goals of our agencies is mandatory. Creating nice-to-know but useless work is a waste of tax dollars and self-sabotage. How will we be able to influence managers with our products? Will we be able to do systematically what many analysts value individually: impact on crime?

THE FUTURE

Problem Analysis[188]

Problem analysis in policing moves beyond traditional crime analysis in that it not only supports policing actions but also drives them. It moves beyond mapping crimes, trend

---

[187]Jerry H. Ratcliffe. See URL *http://jratcliffe.net/research/three_i_model.htm.*
[188]See *Using Analysis for Problem Solving: A Guidebook for Law Enforcement, http:// www.cops.usdoj.gov/mime/open.pdf?Item=342.*

analyses, and pattern identification to using formalized methods of study to arrive at practical solutions.

It is interpretive, creative, and innovative as well as open-ended and inclusive. That is, it does not presuppose the answers to questions or use solely conventional methods and data to examine problems.[189]

Because high-level problem analysis is not being routinely practiced by most police agencies, there are not many examples to demonstrate its relevance and worth or to use as models of integration. Most police agencies do not yet view problem analysis as a necessity to police business and they need to be convinced.[190]

Advancing problem analysis in policing is challenged by the reactive nature of policing and the difficulty of convincing practitioners that problem analysis is a worthwhile effort. However, with a possible increase of crime rates on the horizon, the shift of focus to homeland security, and the fierce competition for resources, it may be an opportune time to assert and adopt the notion of policing "smarter" instead of policing "more."[191]

The conceptualization of problem analysis encompasses a move toward a different sort of policing based on comprehensive analyses, which has not yet manifested itself in reality except in isolated instances. It is not yet integrated into policing.

The first and most obvious reason that problem analysis has lagged behind responding to problems is that, historically, catching the bad guys has been the primary focus of the police, rather than analyzing crime and disorder problems. Police officers and detectives are trained to respond to one call at a time or to investigate one crime at a time. A good detective may link a number of crimes together through a similar perpetrator and/or modus operandi, but this is still examining the crime on an individual level. Thus, policing accumulates and values a different kind of knowledge, that is, experiential knowledge. Research knowledge has not been accumulated or valued as highly. The key is to blend these two types of knowledge as each improves the value of the other.[192]

Problem-Solving Policing is research-based policing. The research is best accomplished internally within the policing agency in order to appreciate the ground-level perspective, and to include all stakeholders. It is reflective rather than reactive, aggregate-based rather than incident-based, and seeks to understand crime as a phenomenon to be addressed in whatever ways may work, with a focus on prevention rather than punishment.

---

[189] COPS, *Problem Analysis in Policing* (Washington, DC: The Police Foundation, March 2003), 3.

[190] *Problem Analysis in Policing*, 34.

[191] *Problem Analysis in Policing*, 45.

[192] Rachel Boba, "Problem Analysis in Policing: An Executive Summary," *Crime Mapping News: Special Issue*, 5, no. 1 (Winter 2003), 2. URL *http://www.policefoundation.org/pdf/vol5issue1_color.pdf.*

Crime Science and Evidence-Based Policing

"Crime science" and "evidence-based policing" undertake a scientific approach to crime much like a doctor applies medical science to his or her patient.

Malcolm Gladwell, in *The Tipping Point*, points out that our notion of criminals as people with evil dispositions who need to be locked up or somehow fixed renders us helpless in preventing crime. We feel that all we can do is defend against such people by locking things up more securely or moving out of high-crime areas. But the reality is, context matters — crimes occur in contexts.[193] Environmental cues, as described in the Broken Windows Theory,[194] contribute to crime — litter, vacant buildings, graffiti — they are the tipping points that affect neighborhoods by encouraging low-level crime.

Professor Gloria Laycock, Director of the Jill Dando Institute of Crime Science, challenges traditional criminology's focus on the deviant criminal and conditions such as poverty and poor parenting that "make" criminals. According to her, evidence in one large study suggests that half of all male offenders are convicted of only one crime in their lifetime, and slightly over half engage in criminal behavior for less than a year.[195] The implications of this concept are enormous. It challenges our assumption that crime is committed only by *chronically* "bad" people. It lends us to accept that a portion of crime is situational and can be prevented — people offend when there is opportunity to do so.[196] Crime science is a practical approach to investigating such contentious concepts, and applying suitable preventive measures.

The UCL Jill Dando Institute of Crime Science is the first in the world devoted specifically to reducing crime. It does this through teaching, research, public policy analysis and by the dissemination of evidence-based information on crime reduction.

Our mission is to change crime policy and practice. The Institute plays a pivotal role in bringing together politicians, scientists, designers and those in the front line of fighting crime to examine patterns in crime, and to find practical methods to disrupt these patterns.[197]

Evidence-based policing requires documentation of outcomes in order to see what "treatments" are effective. Very little assessment of effectiveness of police work in relation to crime prevention is now carried out across the law enforcement system.

The new paradigm of "evidence-based medicine" holds important implications for policing. It suggests that just doing research is not enough and that proactive efforts

---

[193]Malcolm Gladwell, *The Tipping Point: How Little Things Can Make a Big Difference* (New York: Little Brown and Company, 2002), 166-167.

[194]*http://en.wikipedia.org/wiki/Broken_Windows.*

[195]Gloria Laycock, *Launching Crime Science*, November 2003, 3-4. *http://www.jdi.ucl.ac.uk/ downloads/crime_science_series/pdf/LAUNCHING_CS_FINAL.pdf.*

[196]This does not run counter to the theory that 5% of criminals commit 50% of the crime — it refers to the other 50% of the crime, much of which is likely to be preventable.

[197]*http://www.jdi.ucl.ac.uk/.*

are required to push accumulated research evidence into practice through national and community guidelines. These guidelines can then focus in-house evaluations of what works best across agencies, units, victims, and officers. Statistical adjustments for the risk factors shaping crime can provide fair comparisons across police units, including national rankings of police agencies by their crime prevention effectiveness. The example of domestic violence, for which accumulated National Institute of Justice research could lead to evidence-based guidelines, illustrates the way in which agency-based outcomes research could further reduce violence against victims. National pressure to adopt this paradigm could come from agency-ranking studies, but police agency capacity to adopt it will require new data systems creating "medical charts" for crime victims, annual audits of crime reporting systems, and in-house "evidence cops" who document the ongoing patterns and effects of police practices in light of published and in-house research. These analyses can then be integrated into the NYPD COMPSTAT feedback model for management accountability and continuous quality improvement.[198]

The medical concept of diagnosis also applies to understanding neighborhood crime. If we do not ask the neighbors about their condition, how can we assume we have a full understanding of the nature of a neighborhood problem and the best possible "treatment?"

The method we promote to understand neighborhood crime is the method of diagnosis. We are using the term diagnosis here in its broadest sense. To diagnose is to detect, analyze and formulate a prescription for a problem. Research that merely formulates a hypothesis and tests it does not, therefore, fit into this definition. Neither does research that is formulated to prove, or disprove, a theoretical perspective on crime. However, it is quite another matter when those types of research collaborate with those being researched, including decision-makers and practitioners, in order to formulate a practical solution to the crime being studied. That is the method of diagnosis to which we refer.[199]

Crime science, evidence-based policing, and the concept of diagnosis all are rooted in *applied* science, rather than theory.

Stephen Marrin, formerly an analyst at the CIA, also compares medicine with intelligence analysis:

Intelligence agencies should assess medical practices for possible use in improving the accuracy of intelligence analysis and its incorporation into policymaking. In particular, the processes that the medical profession uses to ensure diagnostic accuracy may provide specific models that the intelligence community could use to improve the accuracy of its intelligence analysis. In addition, the medical profession's accumu-

---

[198]Abstract from *Evidence Based Policing*, by Lawrence W. Sherman, in *Ideas in American Policing* (Washington, DC: Police Foundation, July 1998).

[199]Greg Saville and Chuck Genre, draft manuscript, *Understanding Neighborhood Problems* (New Haven, CT: University of New Haven, forthcoming).

lation, organization, and use of information for purposes of decision-making could provide a model for the national security field to adapt in its quest for more effective means of information transfer. While there are some limitations to the analogy due to intrinsic differences between the fields, the study of medicine could provide intelligence practitioners with a valuable source of insight into various reforms that would have the potential to improve the craft of intelligence.[200]

Integration of Neighborhoods into Homeland Security Strategies

Analysis and analysts will likely play a central role in the fuller integration of neighborhoods and other community services into crime intelligence analysis, and — through database development initiatives and liaison with regional, state and appropriate national offices — into homeland security.

Effective homeland security, as well as the provision of public safety services, now and in the future, requires close and continuous coordination and cooperation across the spectrum of social resources and organizations, both geographically and electronically. To be effective this coordination and cooperation must occur at all levels of government (federal, state, and local), among all governmental sectors (police, fire, EMS, transportation, social services, military), between the public and private sectors (corporate security, business, not for profit), and between public servants and public citizens, within the neighborhood and otherwise.[201]

A notable effort toward building a foundation for making this vision a reality comes from Gregory Saville, who proposes a new type of analytical work focused on developing "a cohesive community." His vision is captured in the text box on the following pages.

---

[200]Stephen Marrin and Jonathan D. Clemente, MD, draft manuscript, "Tumors Don't Read Textbooks: Looking to the Medical Profession for Ideas to Improve Intelligence Analysis," e-mail to the author, 13 June 2004. Subsequently published as "Improving Intelligence Analysis by Looking to the Medical Profession," in *International Journal of Intelligence and Counter Intelligence 18*, no. 4 (Winter 2005): 707-729.

[201]Thomas J. Cowper, "Network Centric Policing: Alternative or Augmentation to the Neighborhood-Driven Policing (NDP) Model?" in Carl Jensen III and Bernard H. Levin, *Neighborhood Driven Policing,* Police Futurists International: Proceedings from the Futures Working Group (USDOJ, FBI, Futures Working Group, January 2005), 23. See URL *http://www.fbi.gov/hq/td/fwg/neighborhood/neighborhood-driven-policing.pdf.*

A Vision for Crime Prevention Through Ties That Bind[202]

In recent years following 9/11 there has been great attention focused on security across the country. Most of that has emerged from fears of foreign terrorist attack, fears that for the most part are not founded on real events. With a few tragic exceptions, and one spectacular event in New York, foreign terrorist attacks have been virtually non-existent across the country (even domestic terror events have been relatively rare). Nonetheless, new legislation, a new Department of Homeland Security, and numerous conferences and media reports keep the public eye on terror attack risks.

This public attention has led to wide-ranging policy decisions, most of which focus on the actual event of terror attack (such as target hardening strategies, closed-circuit TVs, and probable-maximum-loss analysis). It has also focused on response procedures in the aftermath (such as catastrophe risk modeling). Unfortunately, planning and analysis for prevention prior to such events has been limited to intelligence gathering by enforcement agencies. For the analyst, these are two unfortunate circumstances: the public demand that the government do something and the obsession with event and post-event planning. These circumstances have created two unique challenges for the crime and intelligence analyst.

First, sound analysis depends on correlative or deductive patterns, and those patterns are inferred by data from previous events. A lack of events means a lack of data, and that is the analyst's nightmare. Second, and linked to the first, there is an enormous dearth in creative strategies for anti-terror protection for pre-event prevention

In such circumstances it is therefore reasonable to draw knowledge from similar fields of inquiry where we can test and apply lessons for homeland security. Intelligence analysts are only now beginning to see this potential. For example, one of the first areas of exploration is probabilistic risk assessment modeling used in disaster planning and insurance risk mitigation. Though in the early stages of development, quantitative risk modeling provides a lens — albeit a blurred one — through which analysts can develop simulations.

---

[202] E-mail to author from Greg Saville, 11 August 2005. For further information on Saville and others involved in problem-oriented policing, see *http://www.popcenter.org/aboutCPOP.html*.

A much more detailed lens exists in the new field of Second-Generation Crime Prevention Through Environmental Design (CPTED). For many years 1st-Generation CPTED has been applied to minimizing crime opportunities in the built environment: improving lighting to improve natural sightlines; enhancing territorial control over properties by legitimate users of that space, and the like. Making it difficult for potential terrorists to drive a car bomb in front of a vulnerable school can be a matter of road design, jersey barriers, and redirecting traffic flows. However, those are short-term security responses.

Analytical strategies for 1st-Generation CPTED can be as simple as conducting CPTED surveys at a particular place. However, 2nd-Generation CPTED deals with the social conditions that encourage positive social interaction, conditions that can empower local residents or employees to surveil a particular locale or identify suspicious activities.

In the 2nd-Generation response arena, the analyst is more interested in identifying the social factors that create a cohesive neighborhood or organization. How can we build the social ties that bind people together, rather than simply rely on what they fear? What factors create positive connections between groups, and what are the correlates influencing and detracting from that cohesiveness? For example, this is the kind of modeling under development by Greg Saville and Nick Bereza of the TRM Group consultants. It is the first of its kind to employ both traditional security and 2nd-generation CPTED. It is these social correlates that lead to more sustainable long-term preventive strategies. A cohesive community mobilized against violent radicals inside, and outside, its realm is much more likely to work together with governments to prevent terror attacks and create safe places.

Mr. Saville's vision of the future resonates with the findings of this research project. More research and more development in the areas of understanding, improving and fostering relationships apply to homeland security, law enforcement, and public safety in general. Safety may be the foundation of strong relationships, rather than the outcome. Community policing rests on the premise that the community is the source of our collective strength and the beneficiary of all levels of government, but it is often not practiced in this spirit.

Signage alerting citizens to report suspicious behavior is one way of including the citizens in the process of homeland security, but Mr. Saville's premise is much broader than mere inclusion of citizens — that strengthening social ties is a greater need to prevent crime and terrorism. This approach appears to remain unrecognized by the evidence of current efforts in homeland security. Using appreciative inquiry as a basis of such an effort herself, the author recognized that focusing on the creation of a vision of safety and the possibility of building new ways of working and thinking, rather than a vision that focuses energy against crime and terror, represents a new paradigm that may not be visible yet except to those who are able to integrate the qualitative into what has been a quantified world.

A VISION OF THE FUTURE FROM THE EXPERTS

If the analysts and experts interviewed for this project could change the field of law enforcement intelligence analysis in any way they could in the next ten or twenty years, this is what they would change:

■ Better-Educated Analysts

Analysts need to be better educated and to have continued educational opportunities: "Learning as a process parallels the analytical process at its best."

■ Real-Time, Quality Data and Quality Analysis

Analysts want timely, quality data so that they can do quality analyses. Better collection, better auditing, better documenting is desirable to reduce turnaround time. "Real-Time Crime Analysis" has appeared in the Ventura County, California, Sheriff's Office: Analysts are listening to police radios while they work and have helped officers make arrests by doing so. In a specific example, an analyst listening in gave patrol officers vehicle plate numbers and a physical description based on knowledge of a neighboring jurisdiction's crime analysis reports; subsequently a serial robber was arrested as he was going back to his vehicle which was left in the lot of the premises he had robbed.

■ Focus on Prevention

Analysts and experts universally desire to see a focus on root causes and proactive analysis to prevent crime rather than react to it.

■ Recognition and Integration of the Analyst into LE

Analysts want recognition, understanding and respect. They want analysis to be integrated into the policing world. Integrating analysis would involve analysts working on every major case. Integration requires training every street cop, every investigator, every agent, and every law enforcement manager to appreciate the value of all information, however minute or obscure, for the purposes of analysis.

■ Standards and Professionalism

Analysts want career paths, education and training developed specifically to enhance the understanding and capacities of crime and intelligence analysis and are eager to see much more literature directly applicable to the field.

■ Technology, Tools, Training

Analysts want automation, powerful computers and broadband technology to support transmissions of data to and from the field. In the future, ortho photography, 3-D imaging, biometrics, collected and applied system-wide for analysis, are desired. The possibilities of exploiting nanotechnology in the future were mentioned by participants — its applications are at this point subject only to the limits of professional imagination.

Police officers carry guns and handcuffs — the things they need least often. They need timely, accurate information at their immediate disposal to make sound decisions. The future of policing involves a number of technological tools. Police could have a high-tech Personal Digital Assistant (PDA) to help them work. A guy urinating in the street could put his thumb on the screen for his print, be identified, the incident recorded, a determination made of whether he is on parole or probation made, all right on the spot.[203]

Some see the analyst as the moderator and interpreter of technological systems as they advance. For example, the computer declares that there might be a crime series underway, as inferred through the application of Artificial Intelligence, and the analyst decides the next step — the analyst prioritizes — using a greater cache of information than those observations already in the system.

Fully integrated data-sharing capabilities and fully linked data-sharing capabilities among local, state and federal agencies are seen as the future of analysis and policing. Experts contributing to this study would like real life to be like "Crime Scene Investigation" (CSI), presently a very popular television program. A number of analysts mentioned CSI in one context or another. Nonetheless, they feel that the media portrayal of law enforcement creates a false picture that interferes with new resource allocation. CSI is very different from law enforcement realities.

■ Intelligence-Led Policing

A number of analysts and experts mentioned ILP as a certainty in the future and yet felt the change would not come to pass soon enough to achieve the greatest good.

■ Working Together

Data sharing, teamwork, and empowering others to analyze are forms of working together desired by analysts. Analysts supporting every major investigation with timelines, maps, and other types of synthesized information are in the actionable future. Analysis taking place before officers engage in a major operation — as well as after — is often not the reality of law enforcement.

Having one security clearance process recognized by all agencies from federal to local level would facilitate working together in the future.

■ Systematic Approaches to Capturing Knowledge

Debriefing criminals, making field assessments, capturing tacit/institutional knowledge, and building dossiers or portfolios on known offenders are widely voiced wishes for the

---

[203] Interview.

future. Already, some observers have identified appropriate questions to guide the systematic development of knowledge:

Sample Offender Interview Questions:

1. Why were specific houses chosen for burglary?

2. Why were particular cars stolen?

3. What did the offender find desirable about the physical condition of the offense location?

4. Why was the offender in a neighborhood other than his or her own?

5. Was the offense planned?

Source: Timothy S. Bynum, *Using Analysis for Problem-Solving: A Guidebook for Law Enforcement* (USDOJ, COPS, 14 September 2001), 37.

■ Nurturing the Analyst

Finally, there is the question of how far training (or experience) can take a given analyst. Any reasonably intelligent individual with the right skills and education can be taught to be an effective analyst. But the truly gifted analyst — the truly gifted athlete, musician, or scientist — is inherently better at his or her job by virtue of inborn talents. In all fields, such individuals are rare. They must be nurtured.[204]

Analysts confided that they would like supportive environments where they can use their individual talents effectively.

■ Predictive Analysis

"For higher-level intelligence analysis, there must be science with accuracy to be predictive. For example, The Internet Crime Complaint Center has 20,000 crime complaints — shouldn't we be able to predict what the 20,001st crime complaint will be with all that data at our disposal?" Improved predictive analysis is anticipated in the future, according to a variety of analysts and experts.

■ The Present in the Future

Clearly, many wishes for the future are already being realized in some small way in the present. This is an interesting result of the research. It corroborates a core assumption of appreciative inquiry — that solutions to central problems already exist within the system.

Next

The next chapter returns to the concept of AI, and addresses the power of story. It proposes a model to carry the present type of inquiry to a higher level.

---

[204] Lowenthal, *Intelligence,* 91.

Chapter 11

HARNESSING THE POWER OF STORY
TO EFFECT CHANGE

Although this project did not apply appreciative inquiry to the fullest dimensions of a group participation approach, in theory, by the interview design and by sharing the findings in this text, the AI process has already begun. *Discovery* in the *high-point* stories uncovered these themes: collaboration, support, discovery, creation, invention, influence, achievement, recognition and impact. These themes can be taken to the "dream" phase of AI to create new metaphors and to inspire accelerated change.

Once you understand the concept of social construction, you realize that action happens in the dialogue. The first question you ask is fateful, and the whole system will turn in that direction. Furthermore, given the heliotropic effect (remember that the sunflower always turns its face to the sun), the system will turn toward its most positive images of itself. Those first interviews *are* the action! Add to that the principle of simultaneity which posits that change is simultaneous with the first questions we ask, and you have a process that recognizes the power of human language to create a reality. We create what we imagine.[205]

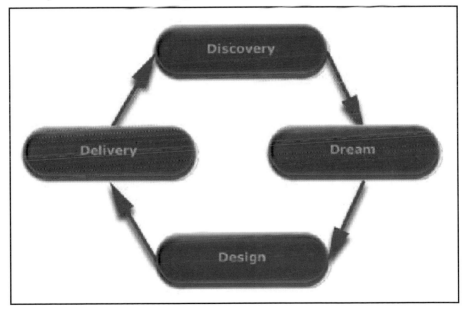

Nodes of the Appreciative Inquiry Process
Source: The AI Cycle. See *http://www.change-management-toolbook.com/tools/AI.html.*

---

[205] Watkins and Mohr, 199.

A number of possible methods exist to use AI on a greater scale if we decide to use this process to address our larger need to collaborate and create change — change in how we collect, share, analyze and use information — what we call *intelligence.*

An AI Summit

If the broad Intelligence Community, including all stakeholders, were to implement an AI process, it might look like this:

- Stakeholders from all agencies would conduct interviews of staff, customers and societal representatives, at all levels of involvement (analysts, officers, agents, commanders, consumers, community, IT staff, academia, policy-makers), keeping interviews tuned as closely to the realities of the work (the data) as possible. The focus of the interviews would be on the experiential: "When we are at our best at intelligence analysis and sharing, what makes work exciting, interesting, invigorating, motivating, and productive?"
- The interviewers, clustered in "synthesis meetings," would then look for the implications, possibilities and inspiration in the data — not necessarily the common ground.
- Meeting outcomes would be shared in small teams which would then create a final report/presentation of all the data.
- A three-day (more or less) *AI Summit* would be held with as many of the stakeholders in attendance as possible. Attendees would determine a variety of organizing possibilities from the findings and agree to a set of design principles to improve the way we analyze and share intelligence, and then determine ways to organize the next steps suggested by the design phase.[206]

What is an "AI" Organizational Summit?

The **WHOLE SYSTEM** participates — a cross-section of as many interested parties as is practical. That means more diversity and less hierarchy than is usual in a working meeting, and a chance for each person to be heard and to learn other ways of looking at the task at hand.

Future scenarios — for an organization, community or issue — are put into **HISTORICAL** and **GLOBAL** perspective. That means thinking globally together before acting locally. This enhances shared understanding and greater commitment to act. It also increases the range of potential actions.

People **SELF-MANAGE** their work, and use **DIALOGUE** — not "problem-solving" — as the main tool. That means helping each other do the tasks and taking responsibility for our perceptions and actions.

---

[206] Cooperider and others, *Appreciative Inquiry Handbook,* 65-66.

COMMON GROUND and NARRATIVE-RICH INTERACTION rather than "conflict management," or negotiation as the frame of reference. That means honoring our differences rather than having to reconcile them, and searching for meanings and direction in stories that honor and connect us to our "history as positive possibility."

APPRECIATIVE INQUIRY (AI) — To **appreciate** means to value — to understand those things of value that are worth valuing. To **inquire** means to study, to ask questions, to search. **AI** is, therefore, a collaborative search to identify and understand the organization's strengths, its potentials, the greatest opportunities, and people's hopes for the future.

INSPIRED ACTION ON BEHALF OF THE WHOLE — Because the "whole system" is involved, it is easier to make more rapid decisions, and to make commitments to action in a public way — in an open way that everyone can support and help make happen. The movement to action is guided by internal inspiration, shared leadership, and voluntary initiative. People work on what they share a passion about, what they most care about and believe will make the difference. Real change begins with the simple act of people acting on what they care about, in the context of a shared vision that matters.

Source: Sarah Patterson and David Cooperrider, Green Mountain Coffee Roasters, *AI Growth Summit Workbook, http://appreciativeinquiry.cwru.edu/ practice/toolsSummitDetail. cfm?coid=5320.*

In Support of *Choosing* New Stories

AI taps into the power of our personal stories focused on what we value (which is always grounded in emotion), and helps us recognize that the possibilities for changes that we need in fact already exist in our work and in the work of others. We can change our stories and define new values in the AI process. We can decide to believe it is possible to create the kind of systems we need to in turn address the threats we perceive to our notions of safety.

> People are not rational. Fact lovers hate this. They want to believe that the "facts are the facts." The story they tell themselves interprets people who aren't rational as an exception rather than a rule. A storyteller embraces, as a central theme, that people aren't rational and uses what she knows about feelings and emotions. She knows that our choices are primarily driven by our feelings. She uses that "fact" to find stories that influence how people feel before she gives them data. Recent studies of how the brain works demonstrate that emotions guide and direct our thoughts and our interpretation of rational facts.
>
> There is ample research to document that decisions are based more on feelings than rational, logical thinking. People decide they like a piece of art because someone they like likes it. They will attribute trustworthiness to an individual they have never met because they have seen his or her picture enough for that person to feel familiar.

They will select one item out of ten identical items and give a list of rational-sounding reasons why it is superior to the other nine — even though the item is exactly the same as the other nine. For each of these feeling-based decisions (they had no facts) research subjects always make up rational-sounding reasons and believe the reasons they make up. People irrationally believe they are rational.[207]

Fear may energize, but the flight or fight energy is from adrenaline, and that adrenaline rush, if continuous, eventually wears out the body. The information overload, the panic, the constant threats we face all *take* energy and leave us in a state of continual crisis. We need to be able to perceive that it is possible to arrive at new solutions and design our future proactively, in an approach rooted in hope, rather than constantly reacting to threats and danger, however real they may be. This does not mean we deny truth, but that we *expand* our vision of the truth and employ all the power we have to create a better future. We can give energy to one another by the power of our stories to motivate, invigorate, and inspire. We can choose new stories.

In *The Springboad: How Storytelling Ignites Action in Knowledge-Era Organizations,* Stephen Denning, former Program Director of Knowledge Management at the World Bank, describes how narrative changed the focus of the World Bank.[208] By using true stories as examples that illustrated the knowledge-management direction he desired for his agency, and helping people add their own visions to possibilities of new stories for the World Bank, Mr. Denning was able to facilitate major change in the information infrastructure and practices at the World Bank.

He found that stories worked better for him than traditional analysis and supporting data and charts in helping people to understand the reason change was needed.[209] As Denning says, "The story does not replace rational analysis, but rather opens up new mental horizons and perspectives, which can then become the subject for analysis and testing." [210]

We Need the Big-Picture Story: Purpose

Change requires that leadership join together to create a bigger vision of what we have yet to achieve — a story of what we are trying to become. AI can help inform leadership in a powerful way and bind stakeholders together in a collective and co-created story.

Systematic structure is the domain of systems thinking and mental models. At this level, leaders are continually helping people see the big picture: how different parts of the organization interact, how different situations parallel one another because of common underlying structures, how local actions have longer-term and broader

---

[207] Annette Simons, *The Story Factor: Inspiration, Influence, and Persuasion through the Art of Storytelling* (Cambridge, MA: Perseus Publishing, 2001), 55-56.

[208] Stephen Denning, *The Springboard: How Storytelling Ignites Action in Knowledge-Era Organizations* (Burlington, MA: Elsevier, 2001).

[209] Denning, 14.

[210] Denning, 177.

impacts than local actors often realize, and why certain operating policies are needed for the system as a whole. But, despite its importance, the level of systematic structure is not enough. By itself, it lacks a sense of purpose. It deals with the *how*, not the *why*.

By focusing on the "purpose story" — the larger explanation of why the organization exists and where it is trying to head — leaders add an additional dimension of meaning. They provide what philosophy calls a "teleological explanation" (from the Greek *telos*, meaning "end" or "purpose") — an understanding of what we are trying to become. When people throughout an organization come to share in a larger sense of purpose, they are united in a common destiny. They have a sense of continuity and identity not achievable in any other way.

Leaders talented at integrating story and systematic structure are rare in my experience. Undoubtedly, this is one of the main reasons that learning organizations are still rare.[211]

Taking Qualitative Inquiry a Step Further...

Is intelligence analysis itself fundamentally qualitative inquiry?[212] If the answer is yes, then would it not be possible to take what is best in AI and apply it to the intelligence analysis process? What might that look like?

Those who collect and analyze information would have their stories (intelligence) gathered as close to the data (all intelligence sources) as possible in the AI model. The resulting summarized intelligence would be clustered in hubs as close to the ground level as possible but highly interconnected in a network. This network could have actual person-to-person relationships to facilitate sharing the intelligence. This type of system would enhance our ability to determine relevancy and exploit our current decentralized information systems. We would not necessarily need all machines (technology) connected, but we would need connections among people. The analysts/hubs could be activated for special needs and have expert access to the information resources, because they would be both intimate with their own data and intelligence and would be highly connected to all the players in their respective regions.

[I]t is from within this matrix of uncertainty, where we are unceasingly crossing the boundaries of established enclaves — appropriating, reflecting, creating — that the vitality of qualitative inquiry is drawn. It is here that we locate the innovative power that is transforming the face of the social sciences. If we can avoid impulses toward elimination, the rage to order, and the desire for unity and singularity, we can anticipate the continued flourishing of qualitative inquiry, full of serendipitous incidents

---

[211] Senge, 354.

[212] Intelligence analysis is fundamentally qualitative because it is looking for specific answers to specific challenges. In many cases, quantitative analysis supports intelligence, but the final intelligence product has no utility if it cannot be acted upon in a qualitative manner.

and generative expressions. In particular, if we bear out the implications of the increasing centrality of relationship over individuals — and realize these implications practically within the emerging spheres of technology — we may effectively participate in the reconstruction of the social sciences and alternation of the trajectories of the cultures in which we participate.[213]

---

[213] Denzin and Lincoln, *Handbook of Qualitative Research*, 1042-1043.

Chapter 12

ARE WE THERE YET?

All intelligence can do is paint a context.[214]

One of the lessons of war is that institutions, while powerful and long-lasting, are often not insuperably rigid if the emergency is great enough.[215]

Crime will evolve and we will evolve.

The CSI Syndrome — the public and therefore Congress thinks we have this stuff.

In the United States we do not look internally with a critical eye at what is or is not working.[216]

When public policing was first formally instituted in London in 1829, the emphasis was on preventing crime: the public and officers themselves regarded successful policing as the "absence of crime."[217]

Analysis as a Responsibility

The 9/11 Commission Report discusses (in the context of the presence of Mihdhar, Hazmi, and Moussaoui in the U.S.), the failure of "the intelligence community to assemble enough of the puzzle pieces gathered by different agencies to make some sense of them and then develop a fully informed joint plan."[218] A very important point follows regarding who was responsible for getting all the pieces together — everyone was responsible.[219] That same concept points to a key failure in the development of systematic analysis in law enforcement: If law enforcement analysis is the responsibility of all officers, then, in reality, no one is responsible. Analysis is a duty. Yet, if few responsible, professional-level analysts are at work in law enforcement (whether dedicated civilian staff or sworn officers detailed to this work), then there may not be enough resources dedicated to carrying out this important duty.

We Are Not Machines

Although analysts must strive to be objective and use technology to "drive" their work, there is a danger in thinking too much in terms of formalizing and facts.

During the past two thousand years the importance of objectivity; the belief that actions are governed by fixed values; the notion that skills can be formalized; and in

---

[214] Quote from interview.

[215] Hayawaka, 175.

[216] Quotes from interviews.

[217] Gene Stephens, "Policing the Future: Law Enforcement's New Challenges," in *The Futurist* 39, no. 2 (March-April 2005): 51.

[218] The 9/11 Commission Report, 355.

[219] The 9/11 Commission Report, 355.

general that one can have a theory of practical activity, have gradually exerted their influence in psychology and social science. People have begun to think of themselves as objects able to fit into the inflexible calculations of disembodied machines: machines for which the human form-of-life must be analyzed into meaningless facts, rather than a field of concern organized by sensory-motor skills. Our risk is not the advent of superintelligent computers, but of subintelligent human beings.[220]

Intelligence analysis, at its basic level, is a very human, practical activity. Existing technology facilitates it. An observer of the national intelligence scene puts it this way:

Change is a constant in the intelligence analysis profession. Technologies change, organizational structures change, and procedures change. Yet the importance of personal relationships is a constant in the production of finished intelligence. In order to access informal knowledge from collectors throughout the Intelligence Community, benefit from the comments of analytic counterparts, and produce analysis relevant to policymakers, personal relationships are crucial.[221]

Risk versus Threat

Police working "smarter," together with federal law enforcers and those whose authorities and responsibilities are to assist — all of us — must work smarter and together. We must realize that the technology on television shows does not exist in most of the real world of policing, that modernity is uneven, technology outdated and/or irrelevant to the work, and that the law enforcement system is fragmented. Nevertheless, our shared mission — to provide safety for the citizens we serve — irrevocably connects us. We are service workers. Networking our resources, developing an educated and trained cadre of team players, being open to innovation and the power of information — these are necessary if we are to maximize our ability to prevent crime and terrorism. If we are to move toward a more secure homeland, analysis, and particularly what we call law enforcement intelligence analysis, will be central to success.

The media tell us that security is "beefed up" when the number of police officers deployed to a given location is increased. But contrary to conventional wisdom, numbers are not a good way to provide security. Rather, it's the ability of law enforcement to recognize suspicious activities and people, their skill at assessing the seriousness of the threat, and the ability to deploy strategies to stop the threat, that minimizes the threat from terrorists who seek to do us serious harm.

The biggest weakness of the current American security approach is the general focus on risk, as opposed to threat. Risk is a calculated assumption made based on past

---

[220] Hubert L. Dreyfus, *What Computers Still Can't Do: A Critique of Artificial Reason*, (Cambridge, MA: The MIT Press, 1999), 280.

[221] Stephen Marrin, *Homeland Security and the Analysis of Foreign Intelligence*, Markle Foundation Task Force on National Security in the Information Age, 15 July 2002, 15. See *http://markletaskforce.org/documents/marrin_071502.pdf*.

occurrences, Threat, on the other hand, is constant. It doesn't matter how many people want to kill you as long as there is one person who wants to kill you. Assuming threat means making it clear to law enforcement that a terrorist attack can happen any day and any time. Having that assumption made translates to a different mode of operations where engagement with a suspect is done immediately and without hesitation.[222]

Capturing timely, accurate data in the field, assessing chronic and emerging problems, being able to note anomalies immediately, and having the ability to link crimes and suspicious activities across jurisdictions require new thinking, not more people.

Police in this war on terrorism will have to be creative, fast thinking, empowered with the ability to make decisions and act decisively in the face of threat and at the same time be defenders of the Constitution and protectors of civil liberties. Achieving this balance involves using these acquired skills that must be constantly tested to insure competence. We need to position ourselves ahead of the enemy, not behind and trying to catch up. In short, we must think like the enemy and act like the defender. [223]

The current law enforcement systems or agencies in this country are generally reactive rather than proactive. They function extremely well in coordinated fashion when faced with obvious threats, such as a sniper case or other murder sprees, but have become numb to many other criminal threats in the constant barrage of "lower"-level, "normal" crime.

The concept of intelligence as used in the military to strategize and maximize forces is only now becoming valued in law enforcement, and, even so, not universally.

Understanding "Common" Intelligence Analysis and Common Threats

Law enforcement intelligence analysis, with its dependence on geographically dispersed eyes and ears, seems to be a logical springboard for the coordination of national, regional and local security resources. It is in light of this understanding that Secretary of the Department of Homeland Security Michael Chertoff urges us to work in a unified manner.

On the most basic level, we need to take a step back and focus on the fundamental question: Why was the Department of Homeland Security created? It was not created merely to bring together different agencies under a single tent. It was created to enable these agencies to secure the homeland through joint, coordinated action. Our challenge is to realize that goal to the greatest extent possible.

Let me tell you about three areas where I plan to focus our efforts to achieve that goal. First, we need to operate under a common picture of the threats that we are facing.

---

[222] Edward J. Tully and E.L. (Bud) Willoughby, "Terrorism: The Role of Local and State Police Agencies," National Executive Institute Associates, Major Cities Chiefs Association, Major County Sheriff's Association, May 2002. URL *http://www.neiassociates.org/state-local.htm.*

[223] Tully and Willowby.

Second, we need to respond actively to these threats with the appropriate policies. Third, we need to execute our various component operations in a unified manner so that when we assess the intelligence and we have decided upon the proper policies, we can carry out our mission in a way that is coordinated across the board.[224]

The Need for Law Enforcement to Change with Changes in Crime

The need for a process-oriented understanding of law enforcement intelligence and its analysis is crucial, as crime will change in the future in ways that we may or may not be able to imagine. A British panel of experts, sponsored by the UK government's Foresight program, considered the theme of crime prevention at the turn of the 21st century. Some results of their thinking about how science and technology can be harnessed for criminal activity as well as by governments for future crime prevention are summarized in the tables below.[225]

We believe new technology now and in the future might also create more opportunity for crime by:

- providing easier access to systems, premises, goods and information;
- removing geographical obstacles to crime;
- increasing the scale of potential rewards; and
- increasing anonymity in committing crime or consuming its proceeds.

The speed and globalisation of criminal innovation versus institutional responses:

Crime thrives in the margins, the gaps where different jurisdictions and regulatory systems meet and conflict — this is likely to allow more opportunity for crime as globalisation increases. Unlike business, it is hard for national law enforcement bodies to have a strong presence elsewhere in the world. Equally the nature of criminal intelligence and the concerns about sharing or losing it do not encourage the breakdown of these divides. The future will be one of crime without boundaries, where more crimes can be committed without the perpetrator ever entering the jurisdiction where the offence impacts. Co-ordinated and concerted proactive steps must be taken with a speed which will match or outpace the criminal.

---

[224] Statement of Secretary Michael Chertoff, U.S. Department of Homeland Security, before the United States House of Representatives Committee on Homeland Security, 13 April 2005, Washington, DC. See URL *http://www.dhs.gov/dhspublic/display?content=4460.*

[225] Department of Trade and Industry, UK Crime Prevention Panel, *Turning the Corner,* 1 December 2000, 9, 11. See URL *http://www.foresight.gov.uk/.* Responses to the original report are at *http://www.foresight.gov.uk/Previous_Rounds/Foresight_1999__2002/Crime_Prevention/ Reports/Turning%20the%20Corner/INDEX.HTM.*

Terrorist Threats and Organized Crime Threats

Assessing the threat of organized crime, especially in light of its links to terrorism, also requires a comprehensive, integrated law enforcement effort grounded in intelligence analysis.

> Both Enron and YBM Magnex, among other similar cases, illustrate some of the potential vulnerabilities of North American business. Greed, coupled with failures in audit and accountability, can combine to form an environment in which organized crime can establish itself and in turn exert real economic influence. And as was the case with Enron, the victim pool may not consist simply of individuals, but could potentially extend to entire financial sectors and their components.
>
> Invariably, incidents like the collapse of Enron and YBM Magnex take us by surprise. They exact a profound toll, not only in terms of their victims, but also with regard to our sense of selves as a civil and transparent society. They also suggest that we do not have a clear sense of the vulnerabilities of our basic economic and financial institutions, particularly with regard to organized crime.
>
> Until we do, meaningful preventative strategies will not be forthcoming and the risks to individuals, groups, and critical sectors will remain acute.[226]

Interview subject Mark R. Gage, Deputy Director for Training and Research for the National White Collar Crime Center, believes that as business is now global, law enforcement must also find ways to globalize. In the past, most business was local. Now, almost no business is local. This has changed the context of crime in ways that law enforcement is ill-equipped to handle. The parochial nature of local law enforcement is outdated.

We need a broader cultural awareness, a sense of the interconnections bigger than our current landscapes, to address the new world with smarter policing. Besides technology and equipment, our policing system itself has not changed much in the last one hundred years.

Homeland Security Strategies Can Incorporate State, Tribal, and Local Public Safety

In November 2004 a project entitled the Taking Command Initiative was launched by the IACP to ascertain what areas of homeland security were working, which were not, and what obstacles stood in the way.[227]

---

[226] Angus Smith, "Crime Victims in a Changing World," in *The Royal Canadian Mounted Police Gazette* 66, no. 3 (2004) online at *http://www.gazette.rcmp.gc.ca/article-en.html?&lang_id=1&article_id=18*.

[227] International Association of Chiefs of Police, *From Hometown Security to Homeland Security: IACP's Principles for a Locally Designed and Nationally Coordinated Homeland Security Strategy,* at *http://www.theiacp.org/leg_policy/HomelandSecurityWP.PDF.*

These law enforcement executives came to the conclusion that our nation's current homeland security strategy is handicapped by a fundamental flaw: *It was developed without sufficiently seeking or incorporating the advice, expertise or consent of public safety organizations at the state, tribal, or local level.*

Further consensus developed over the belief that there was a critical need to develop a new homeland security strategy, one that fully embraces the valuable and central role that must be played by the state, tribal, and local public safety community.[228]

This report goes on to stress five key strategies in the following categories:

- All terrorism is local;
- Prevention is paramount;
- Hometown security is homeland security;
- Homeland security strategies must be coordinated nationally, not federally;
- The importance of bottom-up engineering, incorporating the diversity of state, tribal, and local public safety in noncompetitive collaboration.[229]

A national, rather than a federal, response "ensures that all levels of government, local, tribal, state, and federal are participating in the policy design and development process as *full and equal partners.*"[230] The policies that come from federal agencies are often viewed as overly prescriptive, burdensome, and sometimes impractical.

In fact, the approach taken in the present research, to study those actually doing the kind of work *we say we want more of,* is the opposite of a federal approach. Treating all participants' input as equally valuable is central to Appreciative Inquiry. Prescribing strategies before assessing capacities does not make sense. We need to first understand what is happening on the ground before we "prepare for battle."

Climates of Trust

Building trust within agencies is difficult in law enforcement — building trust with different types of agencies with diverse priorities will be an even greater challenge, one that we cannot hide from. This change will require leadership to develop climates that support the growth of trust.

Law enforcement leadership has an important role in the management of intergroup conflict, as change agents. Management must work toward the development of a culture that is open, willing to share, and trusting. Leadership's inability to achieve such a culture means more and much more of existing problems.

---

[228] *From Hometown Security to Homeland Security, 2.*

[229] *From Hometown Security to Homeland Security, 3-6.*

[230] *From Hometown Security to Homeland Security, 6.*

Diversity accomplished without understanding and vision is similar to red and black ants being thrown together into the same Tupperware container by a grade school child. The accomplishment is remarkable for its ease of implementation and also for its disastrous effect. The change program to diversify the ant colony was quick and easy and short-lived. Change in organizations requires a better plan. Assumptions should not be made on what people think and feel.

Law enforcement must look to better understand and prepare for the change diversity creates in organizations and to develop proactive strategies to manage the resultant conflict. It will, of course, be necessary to establish processes that allow for change implementation, analysis, feedback, and subsequent adjustments. Apart from process will be the necessity of creating a climate or environment that allows for successful change by promoting trust.[231]

As we work toward developing homeland security strategies, it is essential, as an extremely diverse community of law enforcement, national security, and military, that we find new methods to build trust. Trust cannot simply be prescribed.

Changing This Story

Quality analysis is dependent upon quality data, data that can be shared among the stakeholders. We depend on *people* to analyze the data. Users of intelligence depend on intelligent planning of information systems applied by analysts. Effective exploitation of information can bring us toward the objective of crime reduction and homeland security. In that light, for how long can the following news story remain true?

U.S. Info sharing effort off to slow start

By Shaun Waterman

UPI Homeland and National Security Editor
Published 7/28/2005 3:38 PM

WASHINGTON, July 28 (UPI) — The man in charge of the Bush administration's plan to create a permanently connected network of databases to "join the dots" of intelligence against terror threats told lawmakers this week that he had only three people working for him, despite having been in the post more than three months.

John Russack, the program manager for the administration's counter-terrorism Information Sharing Environment, appeared before the Senate Judiciary Committee Wednesday, his first testimony since his appointment by President Bush in April.

---

[231] Lt. Pat Sprecco, "The Importance of Trust in the Management of Bias," *Police Futurist* 11, No. 2 (Fall 2003): 11, 16. See URL *http://www.policefuturists.org/newsletter/pdf/Fall_2003.pdf*

He warned that panel that there were important legal and policy impediments to information sharing that would have to be confronted by Congress and the administration as they moved towards a vision of real-time, seamless data exchange between intelligence, law enforcement, emergency response and other agencies of federal, state, local and other governments.

"Most of the low-hanging fruit has been plucked," he said. "What is left to be done is really hard."

Russack told the panel in his opening statement that he had already begun work. "I will be assisted," he added, "by a very small staff of approximately 25 people," 20 of whom would be "detailees from other parts of our government."

"I'm advised by counsel that you don't have any employees," the committee's Chairman Sen. Arlen Specter, R-Penn., told him.

"Well, I have one," Russack replied, "I have one, and I have two contractors... So we're making progress, Mr. Chairman."

Specter seemed unconvinced. "Sufficient progress?" he asked Lee Hamilton, the former congressman and Sept. 11 commissioner also giving evidence Wednesday.

"It's not even close," replied Hamilton.

Russack earlier said his appointment would run out after two years, but that the law which created the office — last year's intelligence reform legislation — contained "a caveat that says it could actually expire sooner if I don't do a good job."

"Is that sufficient progress, inspector general?" Specter asked the Justice Department's internal watchdog, Glenn Fine, of Russack's hiring record.

"We're going to take a vote here, Mr. Russack," the senator added. You may lose your office sooner."

The exchange came at the end of a marathon four-hour oversight hearing which began focused on the FBI, and hearing testimony from Director Robert Mueller. But with Russack's participation on a second panel, the topic broadened to encompass the challenges facing information sharing across the federal government.

A visibly frustrated Specter asked the other panelists how to instill "a sense of urgency" in the work being done by Mueller and Russack.

"We went to help you," he told Russack. "If I were to write a scathing letter, whom would I address it to, to give you some help?" He asked.

Russack replied that "we have been working hard on this, even though we have a very small staff." He said that he had just sent out a letter to federal agencies and departments outlining the positions he needed filled by detailees.

"I can assure you that there is a sense of urgency to get those positions filled," he said, adding that Specter should write to the Director of National Intelligence John Negroponte.

Hamilton soliloquized about the importance of Russack's job.

"The place where it all comes together is in Mr. Russack's position," he said. "He's the fellow that has to see that we get all this information shared."

Hamilton said the position of the program manager had to be "empowered. He has to have the resources. He's got to have the people. He's got to have the political support in order to get the job done."

Russack said that he had already completed his first legislatively mandated report, delivered to Congress and the president June 15.

His prepared testimony said the report identified five broad issues that would "define the agenda for the program manager's office over its two-year life."

They were — in the order he listed them — "ambiguous and conflicting" policies and authorities governing access to information; a lack of trust between different parts of the federal government, and more generally between the feds and state and local agencies, or between intelligence and law enforcement professionals; the persistence of the need-to-know principle in the application of "controls (on classified data) imposed by the originating organization;" the need to improve information sharing "in parallel with the protection of the information privacy and other rights of Americans;" and, finally, technology.

Russack said that technology was not a barrier to or a restraint on information sharing. "The impediments are not the flow of electrons," he said, calling technology "an enabler of information sharing."

The problem, his prepared testimony explains, is that "disagreements over roles and responsibilities coupled with inadequate or outdated policies, procedures and standards often impede our ability to use existing technology effectively," and resulted in a "vast and confusing array of systems, databases, networks and tools that users must deal with."

Russack's June 15 report is classified For Official Use Only, the level below Secret, he told United Press International after the hearing.

"We'll see what we can do," he responded when asked whether copies might be provided to the media. "A lot of other people have been asking that, too."

Russack told UPI that his office was "going to be staffing up very quickly."

"As you see, I have a lot of oversight to ensure that that happens," he said, but declined to put a timeline on the effort.

Russack was not asked to elaborate on any of the policy or turf conflicts that he alluded to in his prepared testimony. But one such issue identified by the presidential commission on intelligence was the existence of conflicting and inconsistent rules about what intelligence can be collected and shared if it relates to U.S. persons — American citizens and corporations and other people living legally in the United States.

At his confirmation hearing last week, the man nominated to be the general counsel in the office of the director of national intelligence told lawmakers he would work closely with Russack, noting that President Bush had chosen to place him under the new director.

Benjamin Powell said that if confirmed, he "would supply necessary legal support to (Russack) that would involve working with the chief legal officers of the components of the intelligence community to identify legal impediments to information sharing."

Powell said that he was considering setting up "some type of think tank" in the general counsel's office — "people whose job it is to look at these kinds of disputes" and who were "wall(ed)... off from the day-to-day types of tasks that take everyone's time."

We *can* change this story. The effort will require a development of human, professional relationships, collaboration, and vision. With inquiry and dialogue, and perhaps with the articulation of stories rooted in knowledge and an appreciation of what works, change in this direction is a real possibility — among those for whom the desire for change is real.

This study has explored "what works" in law enforcement analysis and what may work in the future. The author applied only the first, Discovery, stage of Appreciative Inquiry. If AI were fully applied to the problems noted in this study, solutions would be designed by the participants and delivered in the Design stage. However, the treatment of all partners as equals in the Discovery stage does not apply to the Design and Delivery stages of the process, where more traditional "command and control" would be applied by managers and political appointees. The goal of this book — to introduce the "outside" world to the work of the analyst in law enforcement — has been achieved through the discovery of valuable truths that frame the world view of active law enforcement analysts.

Appendix

SOME PARTICIPANTS IN THIS STUDY

Participant	Position	E-mail
Thomas Algoe	Retired Customs Intelligence Research Specialist, Buffalo, NY	talgoe@hilbert.edu
Eddie Beach	Field Program Specialist — National Drug Intelligence Center, New York State	Edward.beach@usdoj.gov
Rachel Boba	Professor at Florida Atlantic University in Port St Lucie, FL, Former Crime Analyst	rboba@fau.edu
Lynn Brewer	Crime Analyst in Newport News, VA	lbrewer@nngov.com
Christopher Bruce	Public Safety Analyst in Danvers, MA	Cbruce@mail.danvers-ma.org
Peggy Call	Crime Analyst in Salt Lake City, UT	Peggy.Call@slcgov.com
Tom Casady	Chief of Police in Lincoln, NE	tcasady@lincoln.ne.gov
Howard Clarke	Contract Analyst for RCMP in Vancouver, BC	Howard.clarke@cablelan.net
Ronald V. G. Clarke	Professor at Rutgers in NJ and former head of research in the Home Office, UK	rvgclarke@aol.com
Joseph Concannon	President of InfraGard, Bellerose, NY	Jconcannon@nym-infragard.us
Elise Dekoschak	Intelligence Analyst in the High Intensity Drug Trafficking Area, Tuscon, AZ	edekoschak@dps.state.az.us
Chris Delaney	Research Analyst at Rochester Institute of Technology in Rochester, NY	CD6709@cityofrochester.gov
Shelagh Dorn	Senior Supervisor Program Research Analyst in the Upstate New York Regional Intelligence Center, Latham, NY	sdorn@troopers.state.ny.us

Participant	Position	E-mail
Lisa Palmieri	President of the International Association of Law Enforcement Intelligence Analysts	*Lisa.palmieri@pol.state.ma.us*
Jerry Ratcliffe	Research Professor at Temple University, Philadelphia, PA	*jhr@temple.edu*
Joseph Regali	Manager of New England State Information Network Unit, Regional Intelligence Sharing System	*jregali@nespin.riss.net*
D. Kim Rossmo	Research Professor at Texas State University in San Marcos, TX	*krossmo@austin.rr.com*
Greg Saville	Adjunct Professor in New Haven, CT	*gregsaville@hotmail.com*
Karin Schmerler	Public Safety Analyst in Chula Vista, CA	*kschmerler@chulavistapd.org*
Tess Sherman	Crime Analyst/Investigative Analyst in Austin, TX	*Tess.sherman@ci.austin.tx.us*
Tyrone Skanes	Sworn Divisional Crime Analyst in Toronto, ON	*tyskanes@hotmail.com*
Angus Smith	Senior Policy/Strategic Analyst in the RCMP in Ottawa, ON	*Ac.smith@rcmp-grc.gc.ca*
Mark Stallo	Crime Analysis Supervisor in Dallas, TX	*mstallo@hotmail.com*
Steve Sullivan	Crime Analysis Supervisor in Ventura County Sheriff's Office in Thousand Oaks, CA	*steve.sullivan@ventura.org*
Warren Sweeney	Alpha Group Intelligence Analysis Expert	Warren.sweeney@sympatico.ca
Lorie Velarde	Crime Analyst in Irving, CA	lvelarde@ci.irvine.ca.us

Participant	Position	E-mail
Gregory Volker	Supervisor of Investigations, Kansas City, MO	gvolker@kcpd.org
Susan Wernicke	Crime Analyst in Shawnee, KS	swernicke@ci.shawnee.ks.us
Ronald Wheeler	Intelligence Research Specialist in the US Attorney's Office, Western NY	Ronald.Wheeler@usdoj.gov
Kathleen Woodby	Crime Analysis Supervisor in Austin, TX	Kathleen.woodby@ci.austin.tx.us
Paul Wormeli	Executive Director of the Integrated Justice Information Systems Institute in Alexandria, VA	Paul.wormeli@ijis.org
Robert Ziehm	Operations Intelligence Specialist, U.S. Coast Guard, Buffalo, NY	rziehm@grubuffalo.uscg.mil

Selected Bibliography

Appreciative Inquiry Commons at http://appreciativeinquiry.cwru.edu/.

Barabasi, Albert-Laszlo. *Linked: How Everything is Connected to Everything Else and What It Means for Business, Science, and Everyday Life.* New York: Penguin Group, 2003.

Buslik, Marc. Chicago Police Department, and Michael D. Maltz, University of Illinois at Chicago. *Power to the People: Mapping and Information Sharing in the Chicago Police Department.* URL:http://tigger.uic.edu/~mikem/Pwr2Ppl.PDF.

Bynum, Timothy F. *Using Analysis for Problem Solving: A Guidebook for Law Enforcement.* Washington, DC: U.S. Department of Justice, Office of Community Oriented Policing Services, 2001. URL: http://www.cops.usdoj.gov/pdf/e08011230.pdf.

Cooperrider, David L. and Diana Whitney, *A Positive Revolution in Change: Appreciative Inquiry,* URL: http://appreciativeinquiry.cwru.edu/intro/whatisai.cfm.

Cooperider, David L., Diana Whitney, and Jacqueline M. Stravros. *Appreciative Inquiry Handbook: The First in a Series of AI Workbooks for Leaders of Change.* Bedford Heights, OH: Lakeshore Publishers, 2003.

Dreyfus, Hubert L. *What Computers Still Can't Do: A Critique of Artificial Reason.* Cambridge, MA: The MIT Press, 1999.

Ekblomb, Paul. "Less crime by design." RSA Lectures. 10 November 2000 URL http://www.rsa.org.uk/acrobat/Ekblomb.pdf.

Gale, Stephen. *Standards of Intelligence Reasoning.* Foreign Policy Research Institute: A Catalyst for Ideas, 14 November 2004. URL: http://www.fpri.org.

Gladwell, Malcolm. *The Tipping Point: How Little Things Can Make a Big Difference.* New York: Little Brown and Company, 2002.

Hayakawa, S.I. and Alan R. Hayakawa. *Language in Thought and Action.* Orlando, FL: Harcourt, Inc., 1990.

Heuer, Richards J., Jr. *Psychology of Intelligence Analysis.* Washington, DC: Center for the Study of Intelligence, 1999. URL: http://www.cia.gov/csi/books/19104/art9.html.

International Association of Chiefs of Police. *Criminal Intelligence Sharing: A National Plan for Intelligence Led Policing.* URL: http://www.theiacp.org/documents/pdfs/Publications/intelsharingreport%2Epdf.

_____. *From Hometown Security to Homeland Security: IACP's Principles for a Locally Designed and Nationally Coordinated Homeland Security Strategy.* URL: http://www.theiacp.org/leg_policy/HomelandSecurityWP.PDF.

International Association of Crime Analysts. *Exploring Crime Analysis.* Overland Park, KS: IACA Press, 2004.

Jasparro, Christopher. "Low-level criminality linked to transnational terrorism." *Jane's Intelligence Review,* 1 May 2005.

Klein, Gary. *Sources of Power: How People Make Decisions.* Cambridge, MA: The MIT Press, 2001.

Law Enforcement Intelligence: A Guide for State, Local, and Tribal Law Enforcement Agencies. URL: http://www.cops.usdoj.gov/default.asp?Item=1404.

Laycock, Gloria. *Launching Crime Science.* Jill Dando Institute, November 2003. URL: http://www.jdi.ucl.ac.uk/downloads/crime_science_series/pdf/LAUNCHING_CS_FINAL.pdf.

Lowenthal, Mark M. *Intelligence: From Secrets to Policy.* Washington DC: CQ Press, 2003.

Magruder Watkins, Jane and Bernard Mohr. *Appreciative Inquiry: Change at the Speed of Imagination.* San Francisco, CA: Jossey-Bass/Pfeiffer, 2001.

Major Cities Chiefs Association Intelligence Commanders Conference Report. "Terrorism: The Role of Local and State Police Agencies" in *Terrorism: The Impact on State and Local Law Enforcement,* June 2002. URL: http://www.neiassociates.org/mccintelligencereport.pdf.

National Institute of Justice (NIJ)/International Association of Chiefs of Police (IACP). *Unresolved Problems & Powerful Potentials: Improving Partnerships Between Law Enforcement Leaders and University Based Researchers: Recommendations from the IACP 2003 Roundtable,* August 2004. URL: http://www.theiacp.org/documents/pdfs/Publications/LawEnforcement%2DUniversityPartnership%2Epdf.

O'Shea, Timothy. *Crime Analysis in America: Findings and Recommendations.* Office of Community Oriented Policing Services, U.S. Department of Justice, March 2003. URL: http://www.cops.usdoj.gov/mime/open.pdf?Item=855.

Peterson, Marilyn B., ed., *Intelligence 2000: Revising the Basic Elements.* N.P.: a joint publication of the Law Enforcement Intelligence Unit (L.E.I.U.) and International Association of Law Enforcement Intelligence Analysts (IALEIA), 2000.

Police Foundation. *Problem Analysis in Policing.* COPS: Washington, DC, March 2003.

Rossmo, D. Kim. *Geographic Profiling.* Boca Raton, FL: CRC Press, 1999.

Ryan, Martin A. "The Future Role of a State Intelligence Program." *IALEIA Journal* 12, No. 1 (February 1999).

Sherman, Lawrence W. "Evidence Based Policing." *Ideas in American Policing*. Washington, DC: Police Foundation, July 1998.

Senge, Peter. *The Fifth Discipline: The Art and Practice of The Learning Organization*. New York: Currency Doubleday, 1990.

Simons, Annette. *The Story Factor: Inspiration, Influence, and Persuasion through the Art of Storytelling*. Cambridge, MA: Perseus Publishing, 2001.

Surowiecki, James. *The Wisdom of Crowds*. New York: Anchor Books, 2005.

Thornlow, Christopher C. *Fusing Intelligence with Law Enforcement Information: Police Agencies*. N.P.: National Executive Institute Associates, Major Cities Chiefs Association, May 2002. URL http://www.neiassociates.org/state-local.htm.

U.S. Department of Justice. *Law Enforcement Intelligence: A Guide for State, Local and Tribal Law Enforcement Agencies*. Washington, DC: Office of Community Oriented Policing Services, 2004. URL: http://www.cops.usdoj.gov/Default.asp?Item=1404.

White, Jonathan R., *Defending the Homeland*. Belmont, CA: Wadsworth, 2004.

Whitney, Diana and Amanda Trosten-Bloom. *The Power of Appreciative Inquiry: A Practical Guide to Positive Change*. San Francisco, CA: Berret-Koehler Publishers, Inc, 2003.

Index

A

abilities, 31, 43, 45, 52

academia, 59, 63, 110, 119, 146

access, 3, 7, 8, 9, 11, 33, 40, 57, 58, 60, 63, 72, 82, 88, 90, 96, 129, 149, 152

achievement, 20, 32, 39, 145

analysis, 1, 2, 3, 4, 5, 6, 7, 8, 9, 10, 11, 12, 13, 15, 16, 17, 19, 22, 24, 25, 26, 27, 28, 29, 30, 31, 32, 33, 34, 35, 37, 38, 40, 41, 43, 44, 45, 46, 47, 51, 52, 53, 54, 57, 58, 59, 60, 62, 63, 65, 67, 68, 69, 74, 75, 76

appreciative inquiry, 19, 23, 140, 143, 145, 156

artificial intelligence, 95

artisan, 46, 99

auditing, 141

automation, 87, 141

C

certification, 43

change, 1, 6, 7, 9, 16, 17, 19, 20, 23, 24, 27, 28, 31, 38, 63, 69, 75, 77, 83, 104, 117, 121, 136, 141, 142, 145, 146, 147, 148, 152, 154, 156, 157, 160

characteristics, 5, 31, 43, 84, 94, 104

collaboration, 32, 33, 44, 45, 60, 70, 119, 125, 145, 156, 160

community policing, 67, 68, 69, 75, 82, 100, 115

complexity, 21, 31, 53

COMPSTAT, 67, 68, 75, 76, 86, 100, 137

computer, 34, 38, 53, 63, 67, 77, 81, 82, 83, 88, 95, 98, 133, 142

context, 11, 22, 25, 29, 31, 32, 60, 70, 104, 136, 142, 147, 151, 155

corporacy, 65, 66

creativity 24, 87, 130

crime, 2, 3, 4, 5, 6, 7, 9, 10, 11, 12, 14, 15, 16, 17, 19, 24, 27, 32, 33, 34, 35, 38, 39, 40, 41, 53, 54, 55, 57, 58, 59, 60, 61, 62, 63, 66, 67, 68, 69, 71, 72, 73, 74, 75, 76, 81, 82, 84,

N

National Intelligence Model, 13, 14

national security, 1, 6, 10, 11, 13, 16, 17, 52, 59, 138, 157

networking, 58, 60, 118

networks, 3, 5, 16, 54, 96, 128

O

organization, 1, 19, 20, 21, 22, 24, 29, 30, 35, 38, 48, 67, 76, 77, 84, 96, 98, 100, 112, 113, 120, 122, 123, 125, 126, 138, 140, 146, 147, 148, 149, 156, 157

organized crime, 4, 5, 6, 14, 53, 155

P

paradigms, 21, 22

patterns, 3, 4, 6, 7, 31, 35, 53, 86, 89, 93, 97, 105, 106, 122, 129, 136, 137, 139,

people skills, 48

persistence, 35, 48

planning, 5, 21, 33, 73, 74, 82, 106, 121, 127, 139, 157

positive core, 22, 24

prediction, 32, 41, 99, 122

prevention, 2, 15, 16, 41, 67, 68, 73, 93, 96, 98, 100, 125, 136, 137, 139, 154

principles, 3, 23, 74, 106, 146

proactive, 3, 4, 8, 9, 46, 54, 68, 121, 137, 141, 153, 157

problem analysis, 70, 107, 129, 135

Problem-Oriented Policing, 69, 75

progress, 82, 119, 121

Q

qualitative data, 4

qualitative research, 19, 117

questions, 9, 20, 21, 24, 27, 28, 29, 42, 44, 46, 49, 75, 81, 88, 89, 93, 123, 135, 143, 145, 147

R

recognition, 7, 13, 32, 39, 44, 45, 55, 75, 76, 111, 141, 145

U

understanding, 2, 10, 11, 12, 15, 19, 20, 21, 28, 31, 32, 45, 66, 74, 89, 90, 94, 98, 103, 113, 118, 122, 128, 129, 137, 140, 141, 146, 149, 153, 154, 157

V

vendors, 3, 72, 111

ABOUT THE AUTHOR

Deborah Osborne has been a crime analyst with the Buffalo Police Department for over eight years. She holds a BA in Psychology and an MA in Social Policy with a criminal justice emphasis from Empire State College, SUNY, where she has been an adjunct instructor in Crime and Intelligence Analysis online at the Center for Distance Learning. She has taught this subject at Mercyhurst College Northeast in Pennsylvania and online for Tiffin University in Ohio. Ms. Osborne has been the co-chair of the certification committee of the International Association of Crime Analysts and worked to set standards for a recently offered certification exam. She is also a member of the International Association of Law Enforcement Intelligence Analysts. Ms. Osborne has co-authored a book on crime analysis for practitioners. She has presented at conferences and other venues on the topic of critical thinking skills, including a presentation on "Old and New Thinking Skills to Tackle Organized Crime" at Criminal Intelligence Services Canada National Organized Crime Workshop. Ms. Osborne also has served as an independent consultant to the Police Service of Northern Ireland's Analysis Centre and facilitated short-term placements of PSNI police analysts in law enforcement agencies in North America to promote cross-fertilization of ideas and work methodologies. She is also a member of the International Association for Intelligence Education and Police Futurists International.